Higher GEOGRAPHY

ENVIRONMENTAL INTERACTIONS

The front cover shows the southern part of the Central Business District of the City of Toronto. From centre left to right: the high-rise office towers of the southern CBD, Union Station and railway tracks, the Sky Dome sports stadium with retracting roof and home of the Toronto Blue-Jays, CN Tower, Gardiner Expressway and, on the shores of Lake Ontario, the Harbourfront development.

Editor: Martin Duddin

Morven Archer, Oliver Bray, Jim Bruce

Gordan Lobban, David Russell

HODDER
GIBSON
PART OF HACHETTE LIVRE UK

The Publishers would like to thank the following for permission to reproduce copyright material:

Photo credits
Page 1 © TIM GRAHAM / Alamy; Page 67 MARK EDWARDS / Still Pictures; Page 113 Faras Ghani / Fotolibra; Page 161 © Larry Lee Photography / CORBIS; Fig 1.1 © Peter Hulme; Ecoscene / CORBIS; Fig 1.2 © Robert Estall / CORBIS; Fig 1.5 © Eric Crichton / CORBIS; Fig 1.7 © Doug Houghton / Alamy; Fig 1.8 © PHOTOTAKE Inc. / Alamy; Fig 1.11 © Simmons Aerofilms Ltd / Alamy; Fig 1.12 © Stephen Finn / Alamy; Fig 1.13 ©2004 Woodmansterne / TopFoto; Fig 1.15 © Dr Clifford E. Ford; Fig 1.17 © Skyscan / W Cross; Fig 1.18 www.wikipedia.org; Fig 1.20 © david tipling / Alamy; Fig 1.22 © TIM GRAHAM / Alamy; Fig 1.23 © BMW AG; Fig 1.26 © Jack Sullivan / Alamy; Fig 1.27 Jurassic Coast, Dorset and East Devon World Heritage Site; Fig 1.28 (p.20, clockwise from top left) © AA World Travel Library / Alamy, © Elmtree Images / Alamy, © Paul Thompson Images / Alamy, John Farmar / Ecoscene, © Simmons Aerofilms Ltd / Alamy, (page 21) (left) © Trevor Smithers ARPS / Alamy, (right) © geogphotos / Alamy; Fig 1.29 Adrian Cooper / EASI-Images / CFWimages.com; Fig 1.30 © Photofusion Picture Library / Alamy; Fig 1.31 © Powered by Light / Alan Spencer / Alamy; Fig 1.32 © fstop2 / Alamy; Fig 1.33 LOCOG; Fig 1.34 © AA World Travel Library / Alamy; Fig 1.36 Chris Fairclough / Chris Fairclough Worldwide; Fig 1.37 © B E Eyley / Alamy; Fig 1.38 © Clive Tully / Alamy; Fig 1.44 © Michael John Kielty / CORBIS; Fig 1.45 © Patrick Ward / CORBIS; Fig 1.46 © Adam Woolfitt / CORBIS; Fig 1.47a and b © Robert Pickett / CORBIS; Fig 1.51 (bottom) Tony Waltham Geophotos; Fig 1.54 Ecoscene; Fig 1.62 © Leslie Garland Picture Library / Alamy; Fig 1.68 © Andrew Brown; Ecoscene; Fig 1.70 DYLAN GARCIA / STILL PICTURES; Fig 1.72 (top) David Gowans / Davnik; Fig 1.76 Stock Scotland / Steve Austin; Fig 1.77 Peter Barr; Fig 1.81 © Mike Dodd; Fig 1.83 © John Peter Photography / Alamy; Fig 1.85 Laggan Stores; Fig 1.86 BBC Photo Library; Fig 1.87 © David Robinson / Snap2000 Images / Alamy; Fig 2.08 © Niall Benvie / CORBIS; Fig 2.20 © National Archives / CORBIS; Fig 2.24 © Jim Richardson / CORBIS; Fig 2.25 Franklin D. Roosevelt Library; Fig 2.26 © CORBIS; Fig 2.27 Franklin D. Roosevelt Library; Fig 2.35 (top) © CORBIS, (bottom) © Bettmann / CORBIS; Fig 2.37 Chris Fairclough / Chris Fairclough Worldwide; Fig 2.38 FLPA / Willem Harinck / Holt; Fig 2.39 Nigel Cattlin / FLPA; Fig 2.41 JIM WARK / STILL PICTURES; Fig 2.47 © Wolfgang Kaehler / CORBIS; Fig 2.48 MARK EDWARDS / Still Pictures; Fig 2.50 MARK EDWARDS / Still Pictures; Fig 2.52 J.P.DELOBELLE / Still Pictures; Fig 2.54 MARK EDWARDS / Still Pictures; Fig 3.07 Chris Fairclough / Chris Fairclough Worldwide; Fig 3.14 © Tibor Bognar / Alamy; Fig 3.16 © CanStock Images / Alamy; Fig 3.19 © Wolfgang Kaehler / CORBIS; Fig 3.20 © Ron Watts / CORBIS; Fig 3.26 Ottmar Bierwagen / Spectrum Photofile; Fig 3.27 Jeremy Patterson / Fotolibra; Fig 3.33 Akhtar Soomro / Rex Features; Fig 3.39 © Mark Hakansson / PANOS PICTURES; Fig 3.40 Ilyas J Dean / Rex Features; Fig 3.41 Faras Ghani / Fotolibra; Fig 3.45 AKHTAR SOOMRO / EPA / CORBIS; Fig 3.46 Prodeepta Das / Fotolibra; Fig 4.02 © Rob C. Nunnington; Gallo Images / CORBIS; Fig 4.11 © Larry Lee Photography / CORBIS; Fig 4.13 © Frank Chmura / Alamy; Fig 4.14 © Michael S. Yamashita / CORBIS; Fig 4.16 RON GILING / STILL PICTURES; Fig 4.17 © Sean Sprague / The Image Works / TopFoto; Fig 4.20 Adrian Arbib / Still Pictures; Fig 4.22 © Wolfgang Kaehler / CORBIS; Fig 4.23 © Yann Arthus-Bertrand / CORBIS; Fig 4.24 Robert Harding Picture Library; Fig 4.25 (top) HARMUT SCHWARZBACH / STILL PICTURES, (bottom) MARK EDWARDS / STILL PICTURES; Fig 4.29 World Health Organisation / TDR / Crump; Fig 4.32 © Stephanie Colasanti / CORBIS; Fig 4.33 © Lloyd Cluff / CORBIS.

The publishers would like to thank the Lulworth Estate for permission to reproduce images from the area.

OS map reproduced by permission of Ordnance Survey on behalf of HMSO. © Crown Copyright 2006. All rights reserved. Ordnance Survey Licence number 100036470.

All other photographs supplied by the authors.

Acknowledgements
The authors would like to thank the following individuals and organisations for help given during the writing of this book: Greg Stewart, Program Manager, Transportation Planning, City of Toronto; Angus Murray, Jim Loring and Nick Burn from the Tear Fund; Edwin J Best (Jnr) Librarian, Tennessee Valley Authority; Kerry Watt, Librarian, University of Edinburgh; Eunice Page, Headteacher, West Burton Primary School; Stan Whitaker, Scottish Natural Heritage; Bill Wood, Education Officer, Yorkshire Dales National Park Authority; Dave Parrish, Minerals Planning Officer, Yorkshire Dales National Park Authority; Jim Ferguson, Area Planning Officer, Yorkshire Dales National Park Authority; Robert Moore, Estates Manager, Tilcon Limited; Adekunle Adeoye, Rural and Marine Environment Division, MAFF; South Downs Conservation Board; Countryside Agency; City of Toronto Planning Authority; Isobel Robertson, David Wraight, Wilma Frame, Alan Doherty, Malcolm Macdonald, Dave Orem, Laura Girdwood and Mabel Black.

Every effort has been made to trace all copyright holders, but if any have been inadvertently overlooked the Publishers will be pleased to make the necessary arrangements at the first opportunity.

Hachette's policy is to use papers that are natural, renewable and recyclable products and made from wood grown in sustainable forests. The logging and manufacturing processes are expected to conform to the environmental regulations of the country of origin.

Orders: please contact Bookpoint Ltd, 130 Milton Park, Abingdon, Oxon OX14 4SB. Telephone: (44) 01235 827720. Fax: (44) 01235 400454. Lines are open 9.00–5.00, Monday to Saturday, with a 24-hour message answering service. Visit our website at www.hoddereducation.co.uk. Hodder Gibson can be contacted direct on: Tel: 0141 848 1609; Fax: 0141 889 6315; email: hoddergibson@hodder.co.uk

© Martin Duddin, Morven Archer, Oliver Bray, Jim Bruce, Gordon Lobban, David Russell (2001) 2006
First published in 2001 by
Hodder Gibson, an imprint of Hodder Education,
part of Hachette Livre UK
2a Christie Street
Paisley PA1 1NB
This second edition published 2006

Impression number   5 4 3 2
Year   2010 2009 2008

Cover photo Stone / Getty Images
Illustrations by Peters & Zabransky Ltd
Typeset in Bembo 11/14pt by Fakenham Photosetting Ltd, Fakenham, Norfolk
Printed and bound in Italy

A catalogue record for this title is available from the British Library

ISBN-13: 978-0340-91584-4

Contents

1   **Rural Land Resources** 1

1.1   Rural landscapes in the UK 2

1.2   Coastal landscapes 8

1.3   Carboniferous limestone landscapes 27

1.4   Glaciated landscapes..45

2   **Rural Land Degradation** 67

2.1   Land degradation – What is it? 68

2.2   Land degradation in North America 78

2.3   Land degradation in North Africa 97

3   **Urban Change and its Management** 113

3.1   Population growth and urbanisation in an MEDC 114

3.2   Population growth and urbanisation in an LEDC 141

4   **Development and Health**....................................161

4.1   Development 162

4.2   Health 181

5   **The Examination**
    (Paper 2 Environmental Interactions)...........................195

Index 206

EDITOR'S FOREWORD

This book has been written to cover the requirements of four of the optional six topics for the Environmental Interactions section of the National Qualifications Geography Higher syllabus for the Scottish Qualifications Authority. It is further designed to complement the textbook *Higher Geography: Physical and Human Environments*, also published by Hodder Gibson. This book is designed to replace the original publication *Higher Geography Applications*. Since most students will study no more than two applications, it was decided to select only the four most popular applications for this textbook. Although in part based on the *Higher Geography Applications* textbook, the opportunity has been taken to rewrite large sections of the book and to include newer and relevant case study material essential to further developing the key ideas and concepts introduced through the core. In particular an entirely new section within Rural Land Resources has been written to cover the new topic of Coastal Landscapes. The in-depth case studies, from a variety of Scottish, UK and world locations, as specified in the syllabus, aim to provide students with a variety of resources to allow them to develop skills aimed at:

◆ making use of relevant geographical terminology, concepts and ideas

◆ extracting, interpreting, analysing and presenting geographical evidence

◆ recording information from a variety of sources

◆ developing communication skills in written, map and graphical forms

◆ providing geographical evidence and knowledge in order to encourage independent thinking and the ability to make informed judgements.

Each of the four topics is introduced using clear links to the core content of the Arrangements and should allow seamless integration with core units where this is the preferred teaching method. Text-based questions test students' understanding of key ideas and concepts, and key ideas from the syllabus are summarised at the end of each section. A final chapter gives tips and hints for preparing for the examination and includes examples of exam style questions.

Martin Duddin
February 2006

Section 1
RURAL LAND RESOURCES

Links to the core

This section builds on the ideas and concepts introduced as part of the Lithosphere section of the Physical core unit. It focuses particularly on the formation and characteristic features of three different landscapes within the UK: coasts, upland limestone and glaciated uplands. Links to the Rural Geography unit of the Human core are also developed through case studies of rural landscapes and the pressure for change within the rural resource base of the UK.

1.1 RURAL LANDSCAPES IN THE UK

Rural landscapes provide a variety of physical, economic and social resources. These resources are the product of the interaction between a wide range of physical factors which have been modified by human activity.

Physical factors

The varied rural landscapes of the UK range from the Highlands of Scotland (Figure 1.1) to the East Anglian lowlands (Figure 1.2) and are the result of underlying geology and processes acting upon it (Figure 1.3). Although the UK covers a relatively small area, rock types change considerably over short distances and this results in a variety of distinctive landscapes. As Figure 1.3 indicates, the older and more resistant granitic, **igneous** and **metamorphic rocks** tend to be found in the north and west of the UK.

These are familiar to us as the rocky coastlines and moorlands of south-west England, and the mountain

Figure 1.1 Scottish Highlands

Figure 1.2 East Anglia

Figure 1.3 The geology of the British Isles

massifs of Wales, the English Lake District and Scotland. Elsewhere, particularly in eastern and southern UK, the landscape is based on **sedimentary rocks**, an accumulation of mud, shales, sand and shells formed from pre-existing rocks. These rocks include the upland limestones of the Pennines and the chalk downs of southern England.

It is important to remember that any landscape is not static, but is subject to constant change. Although the UK is not currently subject to dynamic elements of landscape change, such as volcanic eruptions and active glaciers, these have been significant in the formation of the landscape in the past. Long–term processes of landscape change are still at work today, and these are no less significant in altering the land surface. These changes occur so slowly that it is seldom possible to monitor major changes within a human lifetime. The shape of the scenery we see around us is the result of surface processes: the **weathering** and **erosion** of rock, and its subsequent **transportation** and **deposition** (Figure 1.4).

Processes of landscape change

| **Weathering** |
| All processes – mechanical, chemical and biological – by which rocks are loosened and disintegrated by exposure to the elements |

| **Mass movements** |
| Downward movement of weathered material on a slope under the influence of gravity and usually lubricated by rainwater or snowmelt |

| **Erosion** |
| The sculpting of the Earth's surface and scenery by water, wind or ice movement |

| **Transportation** |
| Removal and movement of rock debris by the agents of erosion |

| **Deposition** |
| The laying down of material transported by running water, wind, ice and the sea |

Figure 1.4 The physical processes of landscape change

Although glaciation may have been the most important single factor in shaping the landscape of the Cairngorm mountains in Scotland, surface processes are still at work today modifying the mountain scenery. In much the same way, these processes are also responsible for changing the coastline around the British Isles. Some areas are subject to erosion and coastal land may be lost as a result, whilst in other areas, coasts may be building up due to deposition.

Human activity

Although the geology and surface processes described above have determined the main characteristics of our rural landscape, the human development of the economic potential of the land has been equally important in shaping the present appearance of the rural landscape. Of the economic activities found in rural areas, farming, both arable and pastoral, has had the most significant influence on landscape development, and agriculture remains the predominant land use in rural areas today.

Figure 1.5 Rural Britain

Since the 1851 census, the majority of the UK population have been urban dwellers living in towns and cities. Ever since this rural–urban population shift took place,

> . . . *popular perceptions of rural life have had little in common with reality, focusing instead on idealised cultural images of continuity, stability and changelessness*
> (The National Trust) (see Figure 1.5).

This nostalgic image ignores the reality that rural landscapes are living entities, peopled by individuals and communities.

Our rural and coastal areas are the focus for a number of activities. Perhaps a million walkers, of varying degrees of seriousness, use the hills of the British Isles, not to mention fell-runners, climbers, skiers, sailors, kayakers, canoeists, horse riders, mountain bikers, anglers, paragliders, botanists, sightseers, photographers, trail-bike riders and four-by-four enthusiasts. At the same time rural areas are important to the nation as sources of water, minerals, roofing slate, road-building stone and renewable energy, be it hydro-electric, tidal or wind power.

Rural areas are also important for the maintenance of **biodiversity** and the continued survival of many species of plant and animals, for the production of timber and, controversially, for the training of the military (including the pilots of jets and helicopters). They are the location for large sporting estates based on deer stalking, salmon fishing and grouse shooting. Finally, and perhaps most importantly, they are the homes of local communities who have rights and economic needs, as well as strong cultural roots, complemented in some areas by the Celtic languages, crofting and hill-farming. Not surprisingly, there are frequent clashes and conflicts between the different groups of people using the land.

Every landscape has the potential for change. Some changes may be positive and desirable, others potentially damaging, others still may be both desirable and damaging and result in major conflicts between different interest groups. All of these issues are examined in the following sections of this chapter.

QUESTIONS

1 Using Figures 1.1, 1.2 and 1.3, describe and account for the differences in landscape and relief between the north-west and south-east of the UK.
2 Identify and describe:
   a the main physical processes
   b the main human processes of landscape change in Britain today.
3 How does the nostalgic image of rural Britain ignore the reality?

Land ownership, management and legislation

Virtually all land in the British Isles is privately owned – even within National Parks. Despite an average population density of approximately 250 persons per square kilometre in mainland Britain, most of the 60 million inhabitants live in central and southern England and the central belt of Scotland, leaving much of Scotland, Wales and northern England very thinly populated. A similar population pattern is found in Ireland, with most of the population concentrated in the east, around Greater Dublin and Greater Belfast.

Increased leisure time and greater accessibility have created further pressures on the upland environment of the British Isles. Over the past decade our mountains, moorlands and coasts have become increasingly more accessible due to significant changes in legislation.

Until quite recently, the general public only had a legal right to walk on the various public rights of way – footpaths, bridleways, highways – and on certain registered common land. In England and Wales these rights of way are shown in red on Ordnance Survey 1:50 000 maps and in green on 1:25 000 maps. The situation was changed in England and Wales by the Countryside and Rights of Way (CRoW) Act of 2000. The CRoW Act outlines the statutory right of access on foot to open, uncultivated countryside, defined as mountain, moor, heath, downs and common land. Almost two million hectares of wild land have been enshrined by this law.

The CRoW legislation specifically excludes camping, swimming in non-tidal waters and lighting fires. Landowners can also restrict access to specific areas for up to 28 days in a year for any purpose.

In Scotland the situation was far more complicated. There are almost 7000 rights of way in Scotland but very few are signposted, and they are not shown at all on Ordnance Survey maps. In practice, the public had access to most of the large areas of open land owned by the National Trust for Scotland. Things became clearer in 2003 when the Land Reform (Scotland) Act was considered and approved by the Scottish Parliament.

The Land Reform (Scotland) Act 2003 establishes a statutory right of responsible access to land and inland waters for outdoor recreation, crossing land, and for some educational and commercial purposes.

The Statutory Right of Responsible Access came into force in February 2005. A key part of the Land Reform (Scotland) Act was the development of the Scottish Outdoor Access Code. This code gives detailed guidance on your responsibilities when exercising access rights and if you are managing land and water.

The Act sets out where and when access rights apply. The Code defines how access rights should be exercised. The three key principles for responsible access apply both to the public and to land managers:

◆ **Respect the interests of other people:** be considerate, respect privacy and livelihoods, and the needs of those enjoying the outdoors.

◆ **Care for the environment:** look after the places you visit and enjoy. Care for wildlife and historic sites.

◆ **Take responsibility for your own actions:** the outdoors cannot be made risk-free for people exercising access rights; land managers should act with care for people's safety.

In Northern and Southern Ireland access to rural areas is even more confusing. Farmers are legally entitled to forbid access and there have been a few highly publicised cases of confrontation. Loose dogs may be legally shot if they are seen in the vicinity of farm animals.

Within the Republic of Ireland there are five National Parks, with public access to almost all open land. There are 32 national way-marked trails, totalling over 3300 kilometres, plus many other local trails, all of which are open. However, none of these trails is legally a right of way. A similar situation is found in Northern Ireland. There are few public rights of way in upland areas and landowners are deeply concerned about public liability. However, the Countryside Access and Activities Network (CAAN) is developing a number of way-marked routes and has recommended changes to the law about farmers' liabilities.

Access is a very important issue and it should be noted that a significant by-product of the foot-and-mouth epidemic of 2001 was the wider appreciation of the contribution which tourism – and hill walking in particular – makes to local economies. The epidemic marked the turning point in the relationship between many local councils and leisure groups.

The resultant rural landscapes

The sections which follow look in detail at the evolution, development and current pressures on a

Figure 1.6 Selected rural landscapes of the UK

number of very distinct rural landscape types in different parts of the UK (Figure 1.6):

◆ coastal scenery, focusing on the Dorset Coast, the Firth of Forth and the machair of north-west Scotland

◆ the Carboniferous limestone landscapes of the northern Pennines, with case study material drawn from the Yorkshire Dales National Park

◆ the landscapes resulting from glaciation in the Scottish Highlands, with case study material explaining the rural pressures of the Cairngorm National Park and conflict issues arising from a proposed hydro-electric power scheme in Wester Ross (North West Highlands).

QUESTIONS

4 How does land accessibility legislation differ between Scotland, England and Ireland?

5 Why do you think the foot-and-mouth epidemic of 2001 was important in improving people's understanding of the importance of tourism to the rural economy?

Figure 1.7 Coastal activities: oil production in the North Sea

Figure 1.8 Destructive waves

Scotland's coastline is an astonishing 11 800 km long. That is twice the length of the English coast and longer than the entire eastern coastline of the United States of America; 8% of all the coastlines in Europe are Scottish. (Scottish Natural Heritage, 2002)

The Scottish coastline is one of the most varied and beautiful in the world. Its beaches, cliffs and offshore islands are home to some of the most spectacular gatherings of sea birds and mammals to be found anywhere in the world. But the coast is also a working environment, and many thousands of people depend upon the traditional and new industries of the sea and the coastline. All of these activities, from fishing to oil exploration (Figure 1.7) and tourism, have an effect on the natural environment, and over the past few decades a structure of control measures and legislation has developed to protect the wildlife and its habitats from ill-considered or excessive exploitation of natural resources.

Coastal processes

Waves

Waves are fundamental to the understanding of all coastal scenery. Waves are caused by the wind dragging on the surface of the water. The length of water the wind blows over is called its **fetch**. Strong winds

blowing over a long period of time across a large fetch, will create large waves with high levels of energy.

Out at sea, the waves roll and in a gale they can be over 30 m high. Waves break in shallow water. The water that rushes up a beach is called the **swash** and the water rolling back into the sea is called the **backwash**. There are two main types of wave: **constructive waves** and **destructive waves**. Constructive waves occur when the swash is greater than the backwash; these are the waves that form beaches. Destructive waves (Figure 1.8) occur when backwash is greater than swash; these are the waves that erode beaches. These two types of wave can occur on the same stretch of beach but depend on several factors including wind strength, state of the tide and season of the year.

Erosion

Waves erode in the same way as rivers. They force water into cracks in the rock, under pressure. This helps break up the rock and is called **hydraulic action**. The water dissolves soluble material in the rock; this is called **solution**. Sand, pebbles and large stones scrape against the rock, wearing it away like sandpaper; this is called **abrasion**. Chunks of rock get knocked together and get worn into smaller and smaller bits; this is called **attrition**. Over time this rock ends up as shingle (pebbles) or sand.

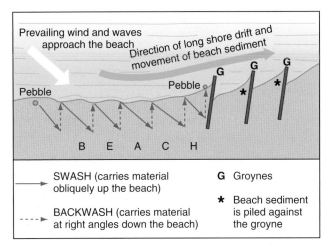

Figure 1.9 Longshore drift

Transportation

The main form of transportation along a beach is **longshore drift**. Longshore drift is the movement of sediment along a beach following the general direction of the prevailing wind. It occurs because the swash of a wave travels up the beach at an angle but the backwash returns down the path of least resistance, which is normally the steepest part of the beach. The swash from the next wave then picks up the same particles and washes them diagonally up the beach again and then the backwash returns the particles down the path of least resistance. The whole process continues, resulting in the migration of sediment along the beach (Figure 1.9).

Deposition

Waves continually carry material on and off the land. If they carry more on than off, a beach forms (constructive waves). Some beaches are made of sand. Some are made of shingle or small pebbles. Waves deposit large particles first, so the beach material is sorted according to size. Figure 1.10 shows the main areas of coastal erosion and coastal deposition in the British Isles.

Figure 1.10 Some areas of coastal erosion (left) and deposition (right) in Great Britain

Geology

The geological structure of the coastline has a considerable effect on the type of coastal features found. If the bands of underlying rock are set at right-angles to the coast, the rivers cut a series of deep valleys, often parallel, to reach the sea. These are known as **discordant coasts** and can be found on the western coast of Scotland.

Concordant coastlines occur when the structure of the geology runs parallel to the coast; concordant coastlines are much less indented than discordant coastlines. The Dorset coastline around Lulworth Cove area (Figure 1.11) is a good example of this, where a parallel band of rock has been breached to allow erosion of softer rock behind.

A third type of coastline is mainly the result of glacial erosion rather than just the rise of sea level. Coastlines such as these have long, very deep inlets from the sea, known as **fiords**. Good examples of this type of coastline are the sea lochs such as Loch Broom in north-west Scotland (Figure 1.12).

Figure 1.12 Loch Broom, north-west Scotland

QUESTIONS

1 Explain why waves are fundamental in the explanation of all coastal scenery.
2 Outline what is meant by longshore drift. Explain its role in the creation of coastal features.
3 How does the underlying geology affect the appearance of the coastline? Use diagrams to help explain your answer.

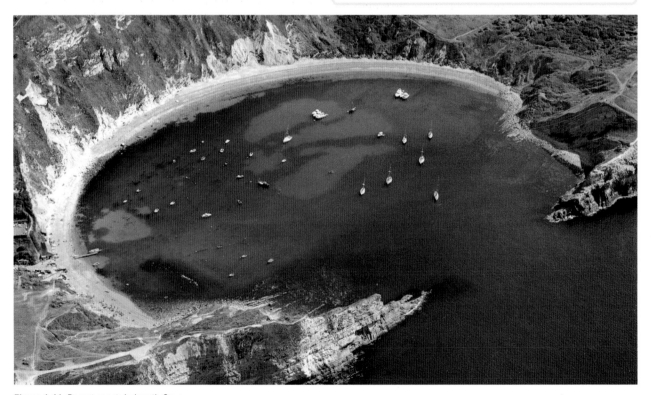

Figure 1.11 Dorset coast: Lulworth Cove

Figure 1.13 The Forth road and rail bridges

The Firth of Forth

The Firth of Forth is the estuary (firth) of the River Forth. It flows into the North Sea between Fife, to the north, and East Lothian, to the south. The river is tidal as far inland as Stirling, but generally it is considered that the inland extent of the firth ends at the Kincardine Bridge. The Kincardine Bridge and

the famous Forth road and rail bridges carry traffic across the Firth (Figure 1.13).

A large number of towns line the shores of the Forth (Figure 1.14) as well as the petrochemical complexes at Grangemouth and Burntisland, the commercial docks at Leith, oilrig construction yards at Dalgety Bay and Methil and the naval dockyard at Rosyth,

Figure 1.14 The Forth estuary

with numerous other industrial areas including the Forth Bridgehead area. The inner Firth (between the Kincardine and Forth bridges) has lost about half of its former inter-tidal area as a result of land being reclaimed, partly for agriculture, but mainly for industry.

The Firth is important for nature conservation. The Firth of Forth islands are Special Protection Areas (SPAs) and are host to over 90 000 breeding seabirds every year. There is a bird observatory on the Isle of May and the Scottish National Sea Bird Centre is in North Berwick.

Geology

From a geological point of view the Firth of Forth and surrounding area is important for a number of reasons. One of these is that it includes a rich geology of rocks from the Ordovician (formed 500 million years ago) to the Carboniferous (formed 280 million to 220 million years ago) period.

It is also the home of Siccar Point (Figure 1.15) and one of Hutton's unconformities. This is a world-famous geological location discovered by James Hutton, who is often considered to be the founder of modern-day geology. Hutton is regarded as the first person to grasp the concept of 'deep time', which is the scientific belief that the Earth is a lot older than biblical interpretation.

The Firth of Forth is designated as a Geological Site of Special Scientific Interest for its contribution to understanding of the Lower Carboniferous geology of the Forth area, and for the worldwide significance of the sedimentary rock sequence for fossil remains. In particular, Wardie Shore is of international importance, having yielded at least eighteen species of fish fossil remains, including sharks, and some of Europe's earliest amphibian remains. The first ever recognised body fossils of the conodont animal, an eel-like fish, were discovered at Granton Shore, the remains of which have been used worldwide in dating rock layer sequences.

The geology of the area has influenced the outstanding coastal landforms which have been modified by coastal erosion, deposition and sea level change.

Coastal features

The coastline of the Firth of Forth is host to a number of spectacular coastal features. These features include a sea arch and stack just south of Thorntonloch (Figure 1.16) and a coastal spit in Belhaven Bay (Figure 1.17).

Figure 1.15 Siccar Point

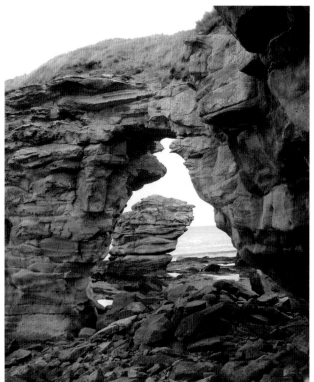

Figure 1.16 Sea arch and stack south of Thorntonloch

Figure 1.17 Belhaven Bay spit

Arches, stacks and stumps form due to differential rates of erosion along a coast that has alternate bands of hard and soft rocks, which results in an irregular coastline made up of bays and headlands. As waves enter the shallow waters of the headland, the influence of the shallow waters causes the wave fronts to bend and be refracted. Due to this wave refraction, wave energy is more concentrated towards headlands and dissipated towards bays. Marine processes – **hydraulic action** and **corrosion** – erode along lines of weakness (for example, joints, faults and bedding planes). Continuous erosion causes back-to-back sea caves to extend backwards until the caves meet and a natural tunnel is formed. The top portion of the sea caves becomes an arch, linking the tip of the headland with the mainland. When the arch collapses, the seaward pillar is left standing and becomes a stack. Continual wave erosion eventually reduces the stack to a stump.

Spits are long, narrow accumulations of sand and/or shingle with one end joined to the mainland and the other projecting out into the sea. They are formed by longshore drift (see Figure 1.9).

Land use

There are a variety of land uses along the Firth of Forth varying from tourism to petrochemicals.

Petrochemicals

Grangemouth (Falkirk) is situated along the southern edge of the Firth of Forth. It has a population of 18 500 and is well known for its large petrochemical plant.

Petrochemical plants, such as the one at Grangemouth (Figure 1.18), are a cause of

Figure 1.18 Petrochemical plant at Grangemouth

environmental concern for many people. They create visual pollution and the many chimney towers and industrial buildings scar the coastal landscape. Barbed-wire fences prevent access from the foreshore. They contribute to air pollution in the form of carbon emissions and water vapour being released into the atmosphere. The large machinery often associated with such works creates noise pollution which can be frustrating for the local residents and marine users. There is also the danger of geothermal pollution (the heating of water due to industry) as sea water is often used for cooling.

However, petrochemicals are important for the country's economy. The industry provides jobs and economic support to local communities. Safety at petrochemical works has improved dramatically in recent years and improved environmental legislation has led to a reduction in emissions and a decrease in geothermal pollution.

Improvements in water pollution

Since 2001 Scottish Water has invested almost £200 million in improving waste-water treatment and cutting sewage outflows in the Firth of Forth. This has included outfalls in Longniddry, Aberlady and Gullane in East Lothian which were completed in 2004 at a cost of £7.5 million. The outfalls move untreated discharges away from beaches and nature reserves, which has led to a significant improvement in the marine environment. Figure 1.14 shows the levels of investment by Scottish Water in making the Forth clean.

A £9 million waste-water treatment works is also to be built in Dunbar in 2006/07 as part of Scottish Water's commitment to clean up the Firth of Forth. The new works will serve domestic and industrial customers in Dunbar, Belhaven and West Barns. Waste water will be treated to a very high level before being discharged into the Biel Water. A storm sewer will also reduce the incidence and impact of storm spills at Belhaven to further protect the quality of the coastal waters.

Investment in cleaner water in the Firth of Forth ultimately benefits all life around the Firth. As well as human activity the Firth is an important conservation area for different types of birds, it is a SSSI (Site of Special Scientific Interest), an SPA (Special Protection Area – a European designation) and a RAMSAR (internationally important wetland) site.

The Forth Estuary Forum – the umbrella group of organisations linked to the Firth of which both Scottish Natural Heritage (SNH) and Scottish Water are members – was set up with the aim of ensuring sustainable use of the Firth. Part of this sustainable use is improving water quality and so promoting biodiversity. It also means cleaner beaches, water and shorelines which residents and visitors can enjoy and be proud of.

QUESTIONS

4 Using diagrams, explain how the stack has been formed at Thorntonloch.

5 Outline the arguments for and against the building of more petrochemical works in the Firth of Forth.

6 Use a table to summarise the information given in Figure 1.14. Why do you think that Scottish Water is so keen to clean up the Forth?

The machair of South Uist

The Scottish Gaelic word 'machair' or 'machar' refers to a fertile low-lying coastal plain. The machair occurs only in the north-west of Britain and Ireland (Figure 1.19), and it is particularly prominent in the Outer Hebrides. There are two different definitions of machair:

◆ a type of sand-dune pasture, subject to agricultural cultivation, which prevails in wet and windy conditions

◆ the span of land from the beaches to the area where sand encroaches on peat bogs further inland.

The machair has received considerable attention from ecologists and conservationists, mainly because of its unique ecosystem. It supports rare carpet flowers, such as Irish lady's tresses, orchids and yellow rattle, along with a diverse array of bird species including the corncrake, twite, dunlin, redshank and ringed plover (Figure 1.20).

Coastal landscapes

Figure 1.19 Distribution of machair

Figure 1.21 Machair system in Scotland

South Uist in the Outer Hebrides has the most extensive cultivated machair system in Scotland (Figure 1.21), extending the whole length of the island and up to 2 km inland. There is extensive grazing, and rotational cultivation, mainly to provide

cattle fodder. The area is very diverse in physical form and is extremely rich in plant species. Extensive areas of wet machair include machair lochs or wet heath, marshes and peatland.

The landscape of the machair is maintained by the interaction between natural processes and the management of croft land. Without active land management by crofters, the landscape of the machair would change. Some of the incentives put in place are helping to retain the human population and conserve some of the traditional agricultural practices.

Figure 1.20 Machair ecosystem

15

Changing land use on the machair

Increasing access to new technology is providing fresh opportunities for land use while rural development and conservation agendas are providing wider opportunities for land management.

The Common Agricultural Policy reforms in 2003 changed the way that farmers were financially supported and some crofters are now looking more creatively at how they can manage their land for agriculture. A good example of this is the Rural Stewardship Scheme which is designed to allow farmers and crofters to farm in a more environmentally friendly way.

The market for tourism and leisure activities on the machair has the potential to grow. Provision of specialist services, such as activity holidays, is likely to provide new opportunities for tourist businesses. There is also a rapidly expanding traditional tourist network of people who come to the Outer Hebrides to sample island life and to enjoy the diverse array of flora and fauna.

There have been conflicts between landowners and tenants over rights to occupy and use land for centuries. Building on previous legislation, the Land Reform Act (Scotland) 2003 radically altered the rights of communities to manage land in Scotland. The legislation has empowered local communities to take control of land that they manage or use.

Renewable energy

The Western Isles also have the potential to be self-sufficient in the production of energy from renewable sources. These renewable energy resources include offshore and onshore wind power and tidal power. Wind power has the potential to provide income to the landowner and the local community, and has been promoted by some as a way of sustaining the local economy. There may also be opportunities to 'export' electricity to the mainland if the national grid infrastructure is upgraded (see page 63).

Windfarms (Figure 1.22) are currently seen by some to be the answer to sustainable rural development and environmentally friendly energy production.

Figure 1.22 A windfarm

There is a continuing debate as to whether it is reasonable to expect one rural area to 'pay' the visual cost of windfarms to satisfy the rest of the country's energy demands. There are a number of proposals to install windfarms on the Western Isles, some of which have caused conflict.

One of the newest proposals was outlined in June 2005, for a hydrogen-powered Hebrides which would put the islands at the cutting edge of renewable energy research and development. A report to the Western Isles Council suggested using electricity from proposed windfarms to produce hydrogen, which can be stored and used to power cars and cut fuel costs. More immediately, it proposes setting up a hydrogen teaching and research laboratory and a hydrogen park to demonstrate new technologies.

Hydrogen is a clean fuel produced from the electrolysis of water and can easily be stored. It can

Figure 1.23 A hydrogen-powered car

then be used, like liquid petroleum gas, as a direct substitute for hydrocarbons, or through powering fuel cells to produce electricity.

Due to their remote location, the Outer Hebrides have some of the highest fuel costs and the highest prevalence of fuel poverty in the UK. With the changing economic climate of energy supply, it is reasonably anticipated that imported fuel costs will continue to rise. Conversely, the Outer Hebrides have one of the richest renewable resources in the world which, if harnessed to produce hydrogen, offers a commercially viable alternative to traditional fuel supply.

The economics of fuel supply in the Outer Hebrides creates a unique opportunity for the early deployment of hydrogen infrastructure for demonstrating the hydrogen economy on a more commercial basis.

The recent proposals also suggest the development of a pipeline to transport hydrogen between Arnish and Stornoway and the provision of filling stations for hydrogen-powered vehicles.

Car manufacturers have already demonstrated hydrogen-driven cars (Figure 1.23), although they need to be mass-produced before being cheap enough to buy. However, in future it is thought that vehicles including motorcycles, buses, ferries, fishing boats and even aircraft could run on hydrogen.

The establishment of a production facility and infrastructure will enable locally produced hydrogen to support research and development, community projects and hydrogen vehicles, and create a new market and jobs. Further economic benefit could come from a community-owned hydrogen production facility.

QUESTIONS

7 Wind power is one of Scotland's most favoured rural resources. Outline why some people are against the building of windfarms.

8 What are the main advantages of using hydrogen created from wind power, rather than just wind power itself?

The impact of severe storms

In recent years the Outer Hebrides have experienced more severe storms, arguably as a result of global warming. In January 2005 parts of the Western Isles were so battered by ferocious storms that the sea may soon be able to break through and link with the Minch, cutting off parts of the islands from each other.

While geologists are fascinated by the possibility of two new Scottish islands being created (Figure 1.24), the issue is more serious for local people who fear that the encroaching seas will damage homes, agricultural land and the delicate machair.

On South Uist the Atlantic Ocean is in danger of breaking through to flood into four freshwater lochs on the west coast of the island. The severe storms in January 2005, which cost the lives of five people on South Uist, tore away chunks of coastline as wide as 50 m, meaning that Loch Altabruig, near the village of Snisheval, is now just 100 m from the ocean. In addition, the loch is separated from the sea by low-lying land, the loch's level is beneath that of the ocean and it is linked to a series of other lochs which in turn link to the Minch. It is feared that the next series of storms will tear away at the strip of land between the loch and the sea.

The land in this area is notoriously prone to erosion because the soil is only a few centimetres deep and overlies a sandy base. Once the sand is exposed, it is

Figure 1.24 Creation of new islands in the Outer Hebrides

quickly blown away, allowing the sea to move further in. The island community have formed an organisation called the Middle District Action Group and are currently campaigning to improve the coastal protection in the area.

Elsewhere in the Outer Hebrides the tiny island of Vatersay, which lies south of Barra, is the subject of similar worries. The north and south parts of the island are linked by a narrow strip of land that was just 500 m wide. The January 2005 storms reduced this strip of land to only 200 m. To the west of this causeway is the Atlantic Ocean and to the east is the Minch.

QUESTIONS

9 If the Atlantic Ocean is able to break through the Minch, what will the social and economic effects be on the local community?

10 Use Figure 1.24 to draw a sketch map of what the Outer Hebrides may look like in 100 years' time.

11 Use an atlas, Ordnance Survey map or the Internet to see if you can locate any other places in the Outer Hebrides where the sea could break through.

The Dorset coastline

Dorset (Figure 1.6) is a county of great beauty, a landscape of contrast and breathtaking scenery. It is one of the maritime counties of southern England, fronting the English Channel which separates the United Kingdom from mainland Europe. Its coastline extends some 140 km from Lyme Regis in the west to Christchurch in the east, and offers a selection of the finest sandy beaches and coastal features to be found in the British Isles. In 2001 the Dorset and East Devon Coast was awarded **World Heritage Site** status for its Jurassic coast.

World Heritage Sites are, by definition: 'places of outstanding universal value recognised as part of the heritage of all mankind'. The Dorset and East Devon Coast World Heritage Site or 'Jurassic Coast' was England's first natural World Heritage Site. There are 812 World Heritage Sites around the world; 628 are of cultural significance, 160 are noted for natural features and 24 are mixed. Great Britain has 4 other natural sites: The Giant's Causeway in Northern Ireland; the island of St Kilda, west of the Outer Hebrides; and two remote islands in the Pacific and Atlantic Oceans. The Dorset and East Devon Coast World Heritage Site is 155 km of unspoilt cliffs and beaches which are easily accessible through 'gateway' towns and the South West Coast Path (Figure 1.25).

Agriculture is the major land use but it accounts for only about 3% of the County's employment. A variety of manufacturing industries are located

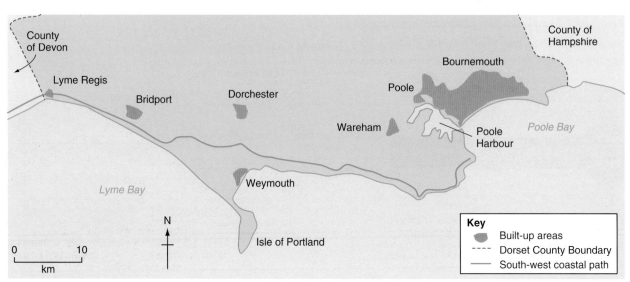

Figure 1.25 The Dorset coast

Figure 1.26 Sailing in Portland Harbour

in Dorset, from old-established traditional industries such as rope and net making (now mainly for military usage) to cutting-edge electronic developments. Tourism is the most important industry, supporting some 25 000 people. Dorset is the third highest ranked county in England in terms of the number of overnights spent by visitors, and second highest for visitor spending.

The importance of the Dorset Coast as an international sailing venue was recently boosted when London was successful in its bid to host the 2012 Olympic Games. The Olympic Sailing (see page 23) will be held at the National Sailing Academy in Portland Harbour which boasts some of the best natural sailing waters in the UK (Figure 1.26).

Management and conservation of the Dorset and East Devon Coast World Heritage Site

A Management Plan has been agreed for the Dorset and East Devon Coast World Heritage Site. This is the key document that establishes the policies for the site and the arrangements for the co-ordinating effort between all the different interest groups who have responsibility for it. The plan clearly sets out proposals for conservation, access, education and science within the area. It also identifies ways in which World Heritage status can help sustainable development in the wider area of Dorset and East Devon.

The future of the site is an exciting long-term commitment for Dorset and East Devon, involving thousands of people who will help to conserve and celebrate the coast. The majority of the work is headed up by a number of working groups. The names, roles and responsibilities of each of these groups are indicated in Figure 1.27.

QUESTION

12 What are the advantages and disadvantages of having so many different organisations and public bodies charged with the task of looking after one piece of coastline?

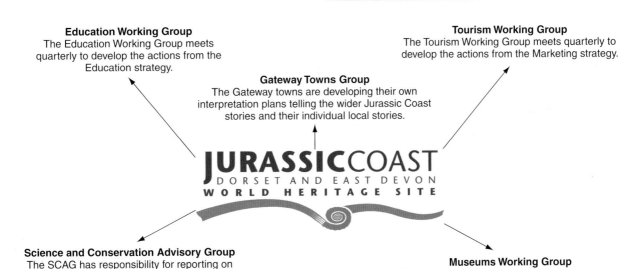

Education Working Group
The Education Working Group meets quarterly to develop the actions from the Education strategy.

Tourism Working Group
The Tourism Working Group meets quarterly to develop the actions from the Marketing strategy.

Gateway Towns Group
The Gateway towns are developing their own interpretation plans telling the wider Jurassic Coast stories and their individual local stories.

JURASSICCOAST
DORSET AND EAST DEVON
WORLD HERITAGE SITE

Science and Conservation Advisory Group
The SCAG has responsibility for reporting on conservation and management of the Site. A Science and Conservation Advisory Network (SCAN) has been established as an additional source of expert advice.

Museums Working Group
The Museums Working Group meets quarterly to plan and discuss Jurassic Coast related plans within the Museum sector.

Figure 1.27 World Heritage Site responsibilities: Dorset and East Devon

Figure 1.28 Dorset coastal features (clockwise from top left): Weymouth Beach, Chesil Beach, Durdle Door Sea Arch, Fleet Lagoon, Lulworth Cove

Figure 1.28 continued: Old Harry Rocks (left), Stair Hole (right)

Coastal features

The Dorset Coast has a wide range of impressive coastal features (Figure 1.28) including Lulworth Cove, the sea arch at Durdle Door, the sand and shingle beach in Weymouth Bay, the cave, arch and stack at Old Harry Rocks, the Fleet Lagoon and Chesil Beach. (The formation of these features is covered comprehensively in the companion textbook *Higher Geography: Physical and Human Environments*, Maclean and Thomson, 2006.)

Tourism and related conflicts in Lulworth

Lulworth is a picturesque village situated on the South West Coast Path in south-east Dorset. It is surrounded by outstanding scenery including Durdle Door, Stair Hole, The Fossil Forest SSSI and Lulworth Cove itself. About 75 000 people visit Lulworth Cove each year; 95% of visitors are day trippers and 35% visit during July and August. Only 10% of annual visits to Lulworth Cove occur between November and February.

The area has a good tourist infrastructure which includes hotels and bed and breakfast accommodation, Durdle Door Caravan Park, Lulworth Youth Hostel and various shops. It is also the home of the Lulworth Heritage Centre (Figure 1.29) which is Dorset's second most visited tourist attraction and its most visited free tourist attraction.

Most people (over 90%) visit the area by car or by coach.

Tourists visiting Lulworth may find themselves in conflict with local people (including the Lulworth estate which is the major local landowner) and the Ministry of Defence.

Local people complain about noise and litter, traffic congestion and inconsiderate parking, particularly during the summer months. There are also a number of unsightly tourist shops which do not fit into the character of the village, and some of these shops raise their prices during the tourist season. Many houses in the area are bought as second homes and lie empty for much of the year. As a consequence, house prices become inflated which means that it is difficult for local people to buy property.

Figure 1.29 Lulworth Heritage Centre

Figure 1.30 Lulworth car park

Some of the traffic problems have been solved by the building of a large car park (Figure 1.30) which is managed by the Lulworth Estate. It can accommodate over 500 cars, and a mini-roundabout has been constructed to allow easy access in and out of the car park. Profits from the car park are used to subsidise a bus service from the local railway station to try and encourage people not to come by car.

Some of the land around Lulworth is used by the Ministry of Defence for live firing (Figure 1.31). Some visitors are unhappy about this because of the noise pollution and the restrictions placed on visitors when roads and the South West Coast Path are closed. The Ministry of Defence has helped resolve this conflict by agreeing a 'no fire' policy on the Lulworth range at weekends and during busy holiday periods, and the roads are also kept open at these times. Firing has also caused damage to the coast. Landslides have occured on the seaward side of Bindon Hill where extensive shell fire has been targeted at the landward side.

The impact of the Olympics on Dorset

The 2012 Summer Olympics, officially known as the Games of the XXX Olympiad, will be held in London from 27 July to 12 August. London will become the first city to host the Olympics three times, having previously done so in 1908 and 1948.

Figure 1.31 Lulworth firing range

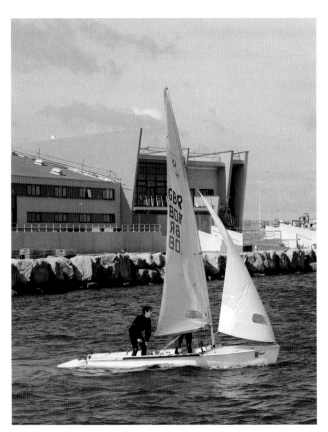

Figure 1.32 Osprey Quay, Weymouth

Figure 1.33 Weymouth and Portland will host the sailing events in 2012

Weymouth and Portland will host the sailing events for the 2012 Olympics. The events will be based around the National Sailing Academy at Osprey Quay (Figure 1.32) on Portland. The National Sailing Academy is built on a site of urban regeneration which was re-developed when the old naval base completely closed down in 1999 due to MOD budgetary cut-backs. The Sailing Academy was officially opened by the Princess Royal in June 2005.

The Olympics will undoubtedly put the Dorset coast in the world's television spotlight as well as bringing sportsmen and women from 200 countries to venues across the south as national teams set up training camps in the run-up to the competitions. Plans for the games are already underway – a cruise liner will be moored nearby to accommodate the competitors.

Due to 50 years of misguided social engineering, the Underhill area of Portland – the backdrop to the Sailing Academy – is desperately in need of a facelift. This is now the chance to give the district, which has been described as one of the most deprived in the South West, a major boost. The former naval residential block will be brought back to life, and all the buildings around Victoria Square will be revamped and redeveloped to give a bright and welcoming gateway to the Isle of Portland. The area

may also benefit from a new supermarket and the building of a new relief by-pass road from Dorchester. This will all bring immeasurable economic and environmental benefits to Portland – the heart of Dorset's World Heritage Jurassic Coast (Figure 1.33).

Local estate agents can barely contain their excitement. Some sellers have upped their prices by £10 000 overnight and shrewd investors have been putting deposits on new harbourside apartments without even seeing them.

The area had a taste of large-scale competition when the World Youth Sailing Championships were held in Weymouth in July 2006. For many local people this was the first test to see if the infrastructure of Dorset can handle the influx of people that large-scale events, like the Olympics, can bring.

QUESTIONS

13 What does the term 'urban redevelopment' mean?
14 'The 2012 Olympics will benefit everyone in the local community'. Explain why this statement is exaggerated.

Coastal defence

Coastal erosion is a huge problem on the Dorset coast. Many of the seaside towns such as Weymouth (Figure 1.34) and Bournemouth rely on their beaches to support the local economy by generating tourism. As a result of coastal erosion, a number of coastal protection systems have been put in place.

In England and Wales no single government department has overall responsibility for coastal protection. It is divided up amongst the Department for Environment, Food and Rural Affairs, the Environment Agency (responsible for protection against river and coastal flooding), maritime local authorities, district councils and county councils, English Nature, the Countryside Commission (responsible for Heritage Coasts), the National Trust, National Parks and National Nature Reserves.

How is the coast protected?

Coasts can be protected by a combination of hard and soft defences. Hard defences involve the construction of major engineering works such as sea walls, groynes and breakwaters. The soft defence approach is when human interference is kept to a minimum and natural processes of beach renewal are maximised. Soft defence methods include beach nourishment schemes and beach stabilisation, for example, reducing the angle of the beach and planting vegetation on areas such as sand dunes.

The soft defence approach: coastal protection and flood alleviation at Preston Beach Road, Weymouth

The main access to and from Weymouth to the east is the A353 (Figure 1.35). It passes between the sea and the wetland RSPB nature reserve at Lodmoor. For approximately 2 km the A353 road is situated just below the level of Weymouth shingle beach which is held back by a sea wall.

Before the 1990s the beach wall was often bombarded by the sea and shingle during winter storms. This frequently led to flooding and the closure of the A353, which in turn led to considerable traffic congestion on Weymouth's other roads.

The cause of the problem has its origin with the direction of tidal flows within Weymouth Bay and the shore wave movements onto Preston Beach which have resulted in longshore drift (Figure 1.9) and the removal of shingle from the beach. To tackle the problem a working group was formed consisting of the Borough Council (as local coast protection authority), the County Council (as Highways Authority) and the National Rivers Authority (which has flood defence responsibilities) working in partnership to implement a new flood alleviation and coastal protection scheme.

Figure 1.34 Weymouth

Figure 1.35 The A353 east of Weymouth © Crown copyright. All rights reserved. Licence number 100036470

Figure 1.36 Soft engineering: beach nourishment

The criterion for the eventual coastal defence was that it must reduce wave energy reaching the foreshore or provide some form of barrier to absorb or withstand storm forces. It was to provide protection without causing detrimental effects to other areas of the coastline. In addition the defence must provide adequate separation between the beach and the road and improve the general amenity of the frontage. The working group considered a number of options including: do nothing, beach nourishment, groyne systems, terminal groynes, offshore breakwaters, submerged reefs, sea walls, and raising the coastal road.

In the end a beach nourishment programme was the preferred scheme (Figure 1.36). Beach nourishment or replenishment is one of the most popular soft engineering techniques. It involves importing sand, gravel or shingle from another area and piling it on top of the existing beach material. The imported sand must be of a similar quality to the existing beach material so that it can integrate with the natural processes occurring there, without causing any adverse effects. Beach nourishment requires constant maintenance and has a 1 to 10 year life before it requires its first major recharge.

In Weymouth shingle was chosen as the material for nourishment. Shingle was chosen because it rests naturally on the foreshore. There were also available supplies of shingle along the south coast of England and shingle beaches satisfy both public and

environmental criteria. The scheme cost approximately £6.5 million to complete and costs the county approximately £250 000 to maintain each year.

The hard defence approach: Swanage

Swanage (Figure 1.37) is a small town in the south-east of Dorset, England. It is situated at the eastern end of the Isle of Purbeck, approximately 10 km south of Poole and 40 km east of Dorchester. The town has a population of 10 124 (2001). The town was a small port and fishing village until the Victorian era when it became a seaside resort. Today, tourism is the main industry in the town. The town has a wooden nineteenth-century pier with two branches; one was refurbished in 1999 and the other is derelict. Sightseeing and angling boats operate from the 'new' pier. Scuba diving takes place under the piers and at some of the nearby coastal wrecks. There is a sandy beach, on Swanage Bay, along the edge of the town. Some of the cliffs to the east of the town are used by climbers. The town is a gateway town for the Jurassic Coast. Like Weymouth, Swanage is reliant on its coastline to support its economy.

A planning application was submitted for a £2.2 million beach regeneration project, which was planned for 2005. The planning application is to build 19 timber groynes along the Swanage shoreline. These would be similar to those found along Bournemouth beach (Figure 1.38) just along the coast from Swanage.

Figure 1.37 Swanage

Figure 1.38 A sea defence groyne at Bournemouth beach

Groynes are wooden, concrete and/or rock barriers or walls built at right-angles to the sea. Beach material builds up against the groyne creating a wider and a more plentiful beach, therefore enhancing the protection for the coast because the sand material filters and absorbs the wave energy. This method also makes the resulting beach more attractive to tourists. It is the most traditional and popular method for coastal protection internationally. However, as the beach material is blocked from moving down the beach by the groynes, this creates serious consequences further down the coast as there are fewer beach particles to move down the coast. Therefore, this creates very rapid coastal erosion and destroys current beaches. Moreover, groynes do not

protect the beach against storm-driven waves and if they are placed too close together will create currents, which will carry sand material offshore.

If no action is taken to safeguard the sea wall at Swanage there is risk in the long term to Shore Road which is one of the main tourist and residential roads in the town. Shore road contains 84 homes, 15 hotels and an extensive subsurface sewer system.

The project will require 120 000 tonnes of sand to regenerate the beach in order to protect the sea wall. Project organisers hope to keep costs down by using sand dredged from nearby Poole Harbour. One of the problems with the construction is that it needs to take place in the winter so that it is not disruptive to tourism.

The government's Department of Environment, Food and Rural Affairs has agreed to fund 80% of the total cost, with the rest funded by council tax over the next 25 years.

QUESTIONS

15 Outline the difference between hard and soft engineering.
16 What are the advantages of beach nourishment over the development of groynes? Are there any disadvantages to the beach nourishment solution?

Figure 1.39 The distribution of Carboniferous limestone in the UK

Carboniferous limestone landscapes occur throughout the UK (Figure 1.39), from Fife in Scotland to the Mendip Hills in south-west England.

The most extensive outcropping of this distinctive rock type is found in the northern Pennines where the limestone forms upland areas which are separated by flat-floored, steep-sided glaciated valleys called **dales**. The Carboniferous limestone found in the Pennines is a hard rock, grey in colour and occurs in massive blocks separated by joints. These joints allow water to pass through them, making the rock **permeable** and resulting in the surface usually being dry, with streams sinking underground.

The processes of erosion and deposition, and particularly the chemical solution of the limestone by rainwater, have produced a distinctive **karst** landscape (Figure 1.40). The role of glaciation processes was important in the development of the present landscape, so it is more correctly called **glaciokarst**.

The Yorkshire Dales National Park

The National Parks of England and Wales were designated by the National Parks and Access to the Countryside Act of 1949. The twin purposes of National Parks were set out in The Environment Act of 1995:

◆ to conserve and enhance the natural beauty, wildlife and cultural heritage of the areas

◆ to promote opportunities for the understanding and enjoyment of the special qualities of those areas by the public.

There are now 13 National Parks in England and Wales (Figure 1.41) of which the Yorkshire Dales is the third largest, covering an area of 1760 km^2

Key	
1 Northumberland	8 Brecon Beacons
2 Lake District	9 Exmoor
3 Yorkshire Dales	10 Dartmoor
4 North York Moors	11 New Forest
5 Peak District	12 South Downs
6 Snowdonia	(designated)
7 Pembrokeshire Coast	13 The Broads

Figure 1.41 National Parks in England and Wales

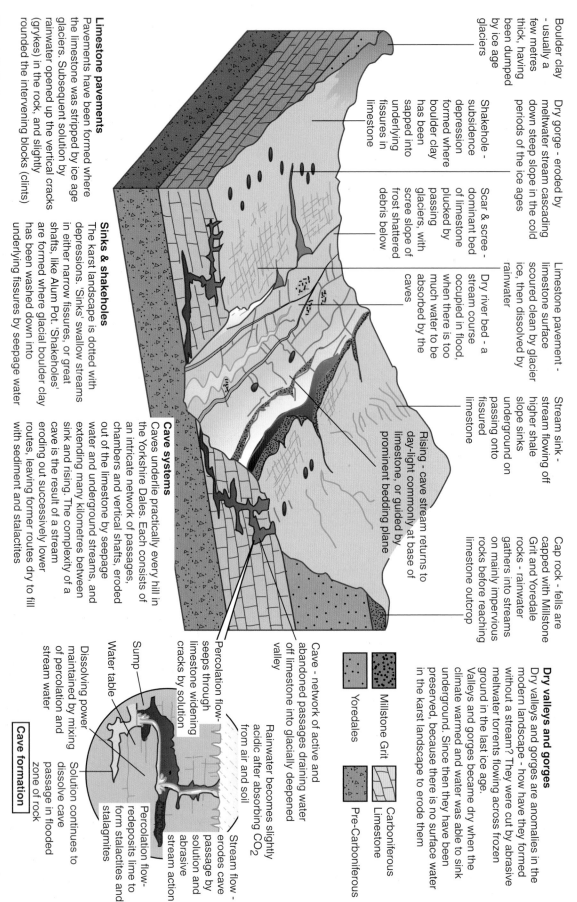

Figure 1.40 Karst scenery

Boulder clay - usually a few metres thick, having been dumped by ice age glaciers

Dry gorge - eroded by meltwater stream cascading down steep slope in the cold periods of the ice ages

Shakehole - subsidence depression formed where boulder clay has been sapped into underlying fissures in limestone

Scar & scree - dominant bed of limestone occupied in stream course plucked by passing glaciers, with scree slope of frost shattered debris below

Limestone pavement - limestone surface scoured clean by glacier ice, then dissolved by rainwater

Dry river bed - a stream course occupied in flood, when there is too much water to be absorbed by the caves

Stream sink - stream flowing off higher shale slope underground on passing onto fissured limestone

Cap rock - fells are capped with Millstone Grit and Yoredale rocks - rainwater gathers into streams on mainly impervious rocks before reaching limestone outcrop

Rising - cave stream returns to day-light commonly at base of limestone, or guided by prominent bedding plane

Dry valleys and gorges
Dry valleys and gorges are anomalies in the modern landscape - how have they formed without a stream? They were cut by abrasive meltwater torrents flowing across frozen ground in the last ice age. Valleys and gorges became dry when the climate warmed and water was able to sink underground. Since then they have been preserved, because there is no surface water in the karst landscape to erode them

Limestone pavements
Pavements have been formed where the limestone was stripped by ice age glaciers. Subsequent solution by rainwater opened up the vertical cracks (grykes) in the rock, and slightly rounded the intervening blocks (clints)

Sinks & shakeholes
The karst landscape is dotted with depressions. 'Sinks' swallow streams in either narrow fissures, or great shafts, like Alum Pot. 'Shakeholes' are formed where glacial boulder clay has been washed down into underlying fissures by seepage water

Cave systems
Caves underlie practically every hill in the Yorkshire Dales. Each consists of an intricate network of passages, chambers and vertical shafts, eroded out of the limestone by seepage water and underground streams, and extending many kilometres between sink and rising. The complexity of a cave is the result of a stream eroding out successively lower routes, leaving former routes dry to fill with sediment and stalactites

Cave - network of active and abandoned passages draining water off limestone into glacially deepened valley

Percolation flow - seeps through limestone widening cracks by solution

Sump

Water table

Rainwater becomes slightly acidic after absorbing CO_2 from air and soil

Stream flow - erodes cave passage by solution and abrasive stream action

Percolation flow - redeposits lime to form stalactites and stalagmites

Dissolving power maintained by mixing of percolation and stream water

Solution continues to dissolve cave passage in flooded zone of rock

Cave formation

Millstone Grit

Yoredales

Carboniferous Limestone

Pre-Carboniferous

Figure 1.42 Yorkshire Dales National Park

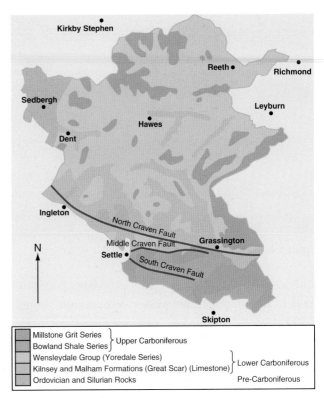

Figure 1.43 Geology of the Yorkshire Dales National Park

(Figure 1.42). The Park has a population of 20 000 people living mainly in small villages. The largest settlement is Sedbergh with a population of 3000. The Park was designated in 1954 primarily to conserve its spectacular Carboniferous limestone scenery, but also the wild moorlands and pastoral valleys (dales) which have given the park its name.

Landscape features and processes

In the area of the Yorkshire Dales National Park, the most impressive features of karst scenery occur in association with the outcrops of **Great Scar Limestone** (Figure 1.43) which was first exposed about two million years ago. This 120–200 m thick slab of rock is almost horizontal and was formed by the gradual deposition of shell debris and chemical precipitates in shallow seas about 330 million years ago during the Carboniferous period. This sediment was compressed over millions of years and recrystallised into limestone which was later cracked and uplifted. The most important process in shaping the limestone scenery has been the repeated climatic changes during the various ice ages.

Two characteristics of the Great Scar Limestone have been crucial to the development of karst scenery. The first is the rock's chemical weakness, which allows it to be completely dissolved by rainwater. Second, in contrast to the chemical weakness, the rock structure itself is mechanically very strong and this has allowed major cave systems to develop.

The amount of rock dissolved by rainwater depends on the acidity levels of the water percolating down through the rock. This is dependent on carbon dioxide levels absorbed from the surface vegetation. This in turn is largely the product of the climate. During warm periods, when the vegetation is plentiful (like the present day), carbon dioxide levels in the water are high and cave systems are dissolved from the limestone. In contrast, during glacial periods ice became the main agent of erosion, with the deepening of the characteristic glacial valleys. In between these alternating periods of warm and cold climate, periglacial conditions prevailed, similar to those found along the edges of the world's tundra regions today. Glacial meltwater is thought to have been responsible for producing some of the most spectacular aspects of limestone scenery, creating **dry valleys** and spectacular **gorges** such as Gordale (Figure 1.44). Initially these landforms were thought to have resulted from the collapse of the roof of underground caverns, but this theory has been

29

Figure 1.44 Gordale Scar

disputed due to the absence of rock debris. It is now widely accepted that the gorges were created during **interglacials**, periods when the ice melted each summer, sending huge volumes of water cascading over the **permafrost** or permanently frozen ground, and causing rapid surface erosion.

Although in some places there is a thin cover of **glacial drift** (the boulder clay deposited over the limestone by glaciers), the soil cover is usually thin, with much of the area now grass-covered moorland. This thin soil cover is easily eroded, and in places the underlying rock has been exposed as a **limestone pavement**, probably the most distinctive landscape feature of glaciokarst (Figure 1.45). Here the well-developed rock joint system of Carboniferous limestone is clearly revealed as percolating rainwater enlarges the joints or **grykes**, which separate blocks of limestone called **clints**. Many limestone pavements were exposed during the last glaciation and have subsequently been eroded and dissolved by rainwater.

Figure 1.45 Limestone pavement in the Yorkshire Dales

Some pavements, for example at Malham Cove, were formed beneath boulder clay. Water, carbon dioxide and bacteria in the soil eroded the underlying limestone, which was then exposed when the ice advanced and removed the thin surface cover. Close examination of Figure 1.45 reveals the presence of small runnels cut into the surface of the clints, which channel rainwater and contribute to the chemical breakdown of the rock surface by solution. In addition to erosion by water and frost action, the pavements are also subject to biological action, because the grykes contain rock fragments and soil which provide a shaded and humid habitat for flora and insects.

Another common feature in areas of karst scenery are enclosed surface depressions known as **dolines**, or in the Yorkshire Dales as **shake holes**. They vary considerably in size, ranging from one to several hundred metres in diameter. They are formed when the thin overburden of soil is slowly flushed down through an open fissure in the underlying limestone, causing the surface to sag and subside. Chemical weathering leads to mass wasting of the sloping surfaces which deepens the hollow. Large areas of Malham Moor are pockmarked with numerous shake holes which have developed in the thin covering of boulder clay deposited by the retreating glaciers after the last ice age.

The harder and more resistant strata within the limestone may be exposed by erosion on the side of hills and appear as prominent cliff-like walls or **limestone scars** (Figure 1.44). Such features are also found underground in limestone areas. Where surface water flows across impervious cap rocks (such as shale) it disappears underground through a **swallow hole** or **sink** where it passes onto the jointed limestone. Stream sinks may be either narrow fissures or huge shafts, but this natural drainage flowing down through the rock creates vast networks of **potholes**, caves and underground passageways which are typical features of the solutional erosion of mountain limestone (Figure 1.40). Almost every hill in the Yorkshire Dales is honeycombed with cave and passageway systems, as rapid stream flow along the horizontal bedding planes and vertical joints in the limestone has eroded out lower levels. The upper caves are often left drier and this has allowed the development of **stalactites** and **stalagmites** through slower percolation flow of lime-rich water through the rock.

The karst scenery of the Yorkshire Dales is of major importance in both a UK and a European context. The processes of weathering and erosion have produced a landscape that is unique in the UK: limestone pavements, cliffs and scars, cave systems and underground rivers. Associated with these features are distinctive grasslands and old woodlands. Over 35% of the limestone dales have been designated as **Sites of Special Scientific Interest (SSSIs)** by **English Nature**, the body which promotes nature conservation in England. As with so many other areas of high landscape value, the Yorkshire Dales face increased pressure from modern agriculture, quarrying and recreation.

QUESTIONS

1   Explain how the processes of glaciation, weathering and erosion have influenced the development of Carboniferous limestone landscapes.
2. List the main landscape features shown in Figure 1.40 and explain their formation.

Rural landscapes and settlement patterns

Changing farms

Farming has transformed the landscape of the Yorkshire Dales. The farmland of the Dales (Figure 1.46) is characterised by an intricate pattern of drystone walls, several hundred isolated stone barns, and flower-rich hay meadows. Livestock farming is the mainstay of the local economy – mainly sheep and beef cattle, with some dairying in the lower dales. The traditional farmed landscape which has resulted is perhaps one of the most distinctive in Western Europe, but has been under threat for some years now due to pressures on farming practices (Figure 1.47a).

Farmers have to respond to changes in their industry, but some aspects of modern farming systems cause concern because of their possible long-term environmental and economic impacts (Figure 1.47b). More intensive production methods have resulted in a loss of traditional farming landscapes and have had a severe impact on flora and birdlife. With changing practices, features such as barns and drystone walls are no longer essential in the farming landscape. Traditionally, cattle were over-wintered

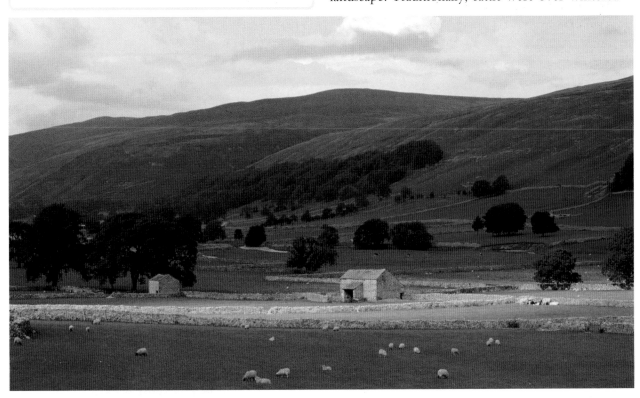

Figure 1.46 Traditional Dales farmland

Increased production costs mean that small farms in marginal areas become less competitive, compared to agri-businesses in lowland areas.

Increasing pressure from tourists.

Sale prices for stock have continued to fall. Increases in cheaper imports, BSE and reduced demand for meat products.

Growth of larger farm units as family farms unable to compete.

Maintenance of 'traditional' Dales landscape becomes uneconomic.

Despite EU grants, subsidies and price guarantees are unable to maintain farmers' income levels.

Figure 1.47a Pressures on farming in the Yorkshire Dales National Park

in these barns, and their accumulated mulch was then spread on the fields as fertiliser during the spring. Hay cropped from the fields was stored in the barns and fed to the cattle during the following winter. The National Park Authority has introduced a scheme to provide funds for the repair of barns and walls in certain areas of the Park. The scheme is jointly funded by English Heritage and the EU.

Some barns have been adapted for use by local craftsmen and others refurbished as visitor accommodation. Some land not suitable for livestock has been given over to new uses such as grouse rearing and farm woodlands. Many Dales farmers have already diversified into providing services for some of the 8.3 million visitor days spent in the Dales each year. Farmhouse accommodation,

European Union Common Agricultural Policy support for cattle and sheep farmers.

Farm diversification – increased income from tourism, farm shops and craft workshops.

UK Government payments for farmers in less favoured areas.

Barns and walls conservation scheme payments.

Environmentally Sensitive Areas (ESAs). Parts of the area are inside the Pennine Dales ESA allowing payments to farmers for reducing intensity of production and promoting conservation.

Countryside Stewardship Scheme (CSS) pays grants to farmers who manage their land in a traditional way.

Figure 1.47b Responses to pressures

campsites and farm shops (Figure 1.48) have been developed to provide additional income, although much of this is seasonal.

The economy of rural areas, once strongly linked to agriculture, is also experiencing great change. Links between farmers and the local community have been weakened as mechanisation has reduced the need for local labour. This has reduced employment opportunities in rural areas and alternative jobs are often difficult to find.

The Yorkshire Dales National Park has the UK's highest concentration of herb-rich meadowland, which is generally regarded as being the habitat most vulnerable to agricultural change. Estimates suggest that 95% of hay meadows in England and Wales have been wiped out since the Second World War.

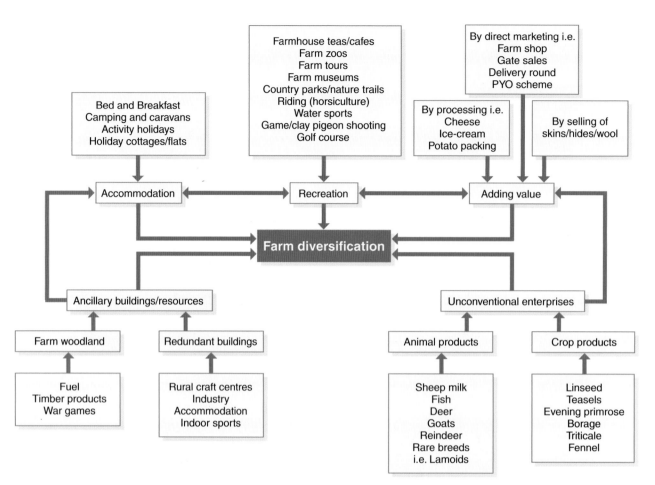

Figure 1.48 Types of farm diversification

Many of those that remain are now protected as SSSIs. In 1987 the Pennine Dales was designated an **Environmentally Sensitive Area (ESA)** (Figure 1.49), covering 16% of the National Park. Under this scheme five-year agreements are offered to farmers willing to manage their land in a traditional way. In return for a payment per hectare, participating farmers agree to maintain existing walls and barns, not to plough up grassland, and to keep stock off meadows for agreed periods. The use of fertilisers, pesticides and herbicides is also strictly controlled.

The changing village

The basic pattern of settlement in the Yorkshire Dales was established by the Danish farmers who colonised the main valleys between the seventh and tenth centuries. By the time of the Domesday Survey of 1086 there were few nucleated villages, but Norman society encouraged village growth, and the present pattern of nucleated villages was well established by the Middle Ages.

These villages, built of local material and in a traditional local style, are fundamental to the Dales landscape. Throughout the UK, villages were largely created to fulfil an agricultural function which has become increasingly less relevant. As the agricultural labour force has declined, and with it the population of rural areas, village services have followed this decline. Over the years, villages have lost their schools, post offices, village shops and cottage hospitals, leaving people to travel long distances for even the most basic services. A Survey of Rural Services was carried out in 1997 for the parliamentary Select Committee on Agriculture. This survey revealed that for many of the basic services required for the conduct of normal daily life, rural parishes were often very much worse off than even the most deprived urban areas (Table 1.1).

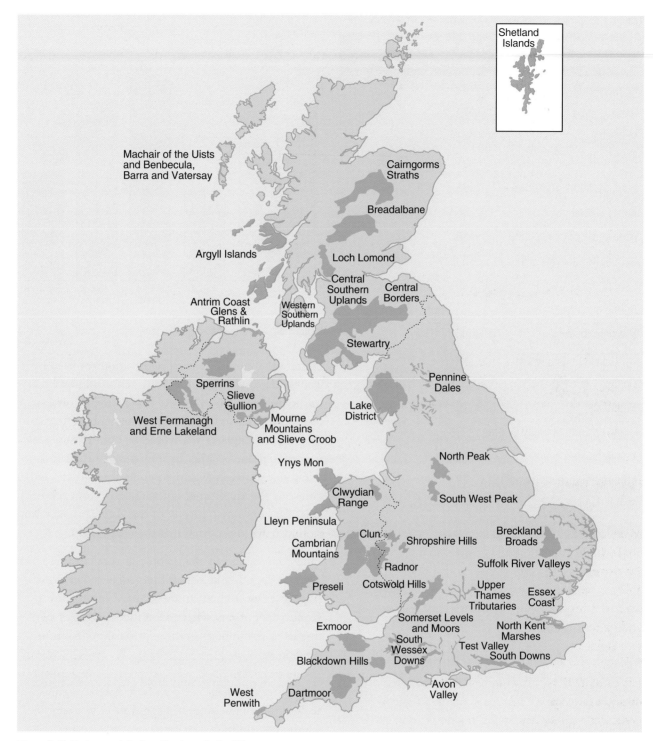

Figure 1.49 Environmentally Sensitive Areas in the UK

Table 1.1 Proportion of rural parishes in England without services (1997)

Service	% of rural parishes without service
Permanent shop (of any kind)	42
General store	70
Post office	43
Bank or building society	91
Bank cashpoint machine	95
Petrol station	56
Newsagent or confectioner	86
School (for six year old)	50
School (for any age)	49
Public nursery	93
Private nursery	86
GP (based in the parish)	83
Dentist	91
Pharmacy	79
Village hall/community centre	28
Public house	29
Library (permanent or mobile)	12
Sports field	50
Youth club	68
Daycare group for the elderly	91
Residential care for the elderly	80
Job centre	99
Official jobs noticeboard	98
Benefits Agency office	99
Citizens Advice Bureau	94
Daily bus service	75
Community minibus/social car scheme	79
Rail service	93
Police station	92
Public telephone	9
Mains sewerage	20
Mains gas	54

Source: 1997 Survey of Rural Services, Select Committee on Agriculture

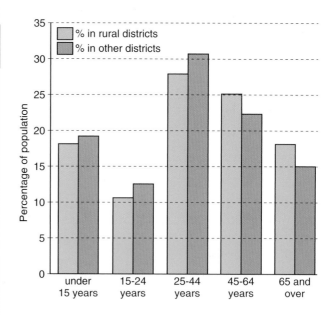

Figure 1.50 Rural population age structure

Despite the picture of rural deprivation revealed by Table 1.1, results from the 2001 census indicate that in the previous ten years, the population of rural districts in England grew appreciably faster (10.3%) than that of England as a whole (5.3%). This increase is primarily due to a net in-migration to rural areas, mainly of relatively wealthy people from towns and cities. At the same time, out-migration from rural areas has continued, mostly of the younger generation to find work and homes that they can afford.

One consequence is that people in rural areas are, on average, slightly older than their urban counterparts (Figure 1.50). Recent studies carried out for the Countryside Agency have identified a number of threats to community spirit and social cohesion in rural areas:

◆ increased mobility and rapid population change

◆ more people commuting instead of working in the community

◆ decline of local shared facilities such as shops, schools and pubs

◆ loss of distinctive rural culture, language and dialects

◆ conflicts between incomers, local people and farmers

◆ loss of younger people.

Rural communities within the Yorkshire Dales National Park exemplify many of these problems. Wensleydale (Figure 1.42) in the north-eastern part of the Park exhibits such symptoms of rural deprivation: an ageing population, lack of employment, lack of housing opportunities for local people, and declining village services. In an attempt to address these problems, the National Park Authority has established a number of initiatives. In Hawes, a small industrial site was established in association with the former Rural Development Commission (now part of the Countryside Agency) and English Estates. Seven small industrial units were built on a local authority site in the mid-1980s and by 2000 this had increased to more than ten units.

Although small-scale, these firms can help to offset the continuing decline of farm-related employment, allowing local people to work close to where they live. With the continuing trend towards **footloose industries** (without locational ties) and high-technology based light industry, rural areas offer many advantages to potential employers, including a stable workforce and a good quality of life.

Provision of jobs and services alone will not solve the problems in rural areas. Low-cost housing is also vital if local families are to be encouraged to stay in the villages. West Burton (Figure 1.51) near Aysgarth, is in an upland area and was originally built around a ford and a crossroads with the local

economy based on lead mining and agriculture. In a recent survey almost half the village population of 270 were in the 'dependent' category, with 60 aged under 16 and 70 of pensionable age. The village is reasonably well provided with services. There is a post office, village shop, butcher's shop, mobile library (which calls every three weeks) and a weekly visit from a fish and chip van. The nearest dispensing doctor's surgery is in Aysgarth, 3 km away, although there are regular prescription deliveries to the local post office. There is a pub, and the village primary school has 34 pupils travelling in from a wide area around West Burton. Buses run most days to both Leyburn and Hawes and there is an informal car-sharing network.

Most of the working population commute to jobs beyond the village, although up to 20 work on family farms and about ten in local service industries. A number of jobs are tourism-related – that is, they are part-time and seasonal in nature.

The National Park Authority is particularly concerned that a substantial number of the area's housing stock is not occupied full-time. Of the 90 houses in West Burton, 32 are second or holiday homes. Eleven houses are privately rented and permanently occupied, and both of the original council houses were sold to their tenants. The local community need more affordable housing to be built in the village.

Figure 1.51 West Burton in Wensleydale

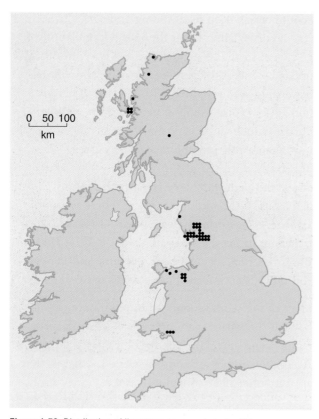

Figure 1.52 Distribution of limestone pavements in the UK

Conflicts in the park

Quarrying in the Yorkshire Dales

The landscape character of the Yorkshire Dales is largely the result of the underlying geology (Figure 1.43). In the same way that this landscape is a valuable resource for the development of tourism, so the rocks and minerals are valuable raw materials. One of the problems for the National Park Authority is to attempt to reconcile the fact that whilst they are charged with protecting the distinctive Dales landscape, they also have to ensure that this aim is balanced with the employment opportunities which activities such as quarrying provide for local communities. Quarrying has traditionally been an important source of local employment and whilst modern large-scale mineral extraction employs fewer people, it is estimated that the quarrying industry provides employment for 7% of the Dales' working population, as well as contributing almost £6 million a year to the local economy.

At **Ingleton** quarry extensive landscaping work has recently been carried out. The quarry plant has been re-sited, screening banks have been built and trees planted. A final restoration scheme based around a central lake has been approved

At **Horton** quarry work continues under extensive permission granted between 1948 and 1968. The site has large permitted reserves and is very prominent in views from the surrounding countryside. Landscaping works and other improvements have been agreed with the quarry company

Arcow quarry at Helwith Bridge is now operating under permission granted in 1986. Major landscaping work has been undertaken, including the construction of a screen embankment around the plant and stock piles

Extraction at **Giggleswick** quarry has reached the outer boundaries of its 1973 permission. Improvements to the processing plant were approved by the National Park in 1993

Threshfield quarry is worked under extensive permission dating from 1952 and there are large remaining reserves. The upper quarry faces are becoming more conspicuous from the surrounding countryside and landscaping works are being carried out to improve the appearance of the quarry

Rail facilities at **Swinden** quarry near Cracoe have recently been upgraded and a new train has been introduced. This will take up to 27800 lorries off the road each year. Permission was granted in 1996 to deepen the quarry by 100m and to extend the life of the quarry to 2020. The site will then be restored and managed for a nature conservation after use

Figure 1.53 Working quarries in the Yorkshire Dales National Park

At present two main rocks are quarried within the National Park: **gritstone** and limestone. Gritstone is derived from Ordovician and Silurian rocks, mainly siltstone and sandstone, and is used for a variety of surfacing operations, notably roads and airport runways, because of its ability to resist skidding. Carboniferous limestone has a high content of calcium carbonate and is used for agricultural and industrial lime, as flux for the steel industry, for glass making and for making toothpaste.

The Yorkshire Dales National Park has the greatest concentration of limestone pavement in the UK (Figure 1.52), with the areas around Malham and Arncliffe providing the most extensive exposures. Limestone in this area has proved a popular resource since prehistoric times when clints were removed and used to build stone circles. During the medieval period, intensive grazing by animals caused soil erosion and exposed further areas of limestone which was burned in the many lime kilns found throughout the Dales. The resulting ash was used to improve the poor grazing land. Later, during the enclosure periods of the eighteenth and nineteenth centuries, the stone was used for walls to separate the fields. Recently, a new threat has emerged with increased demand for limestone as a result of the flue–gas desulphurisation programme of the electricity generating authorities.

Figure 1.53 shows the major quarries still working within the National Park area. The concentration of quarrying activity in the south-west part of the Park on the Kilnsey and Malham or Great Scar Limestone (Figure 1.43) is due to the rock's high chemical purity, with calcium carbonate content generally in excess of 95%. The limestone is several hundred metres thick, and it is relatively easy to quarry. There are fewer quarries today than in the past, but those that remain are much larger and the environmental issues associated with their operation are considerable. Their size creates a visual intrusion on the landscape (Figure 1.54) because the very scale of operation makes concealment impossible. Continued activity also threatens landscape features such as cave systems and wildlife. Blasting, crushing and screening of rock, the loading and burning of lime and transport to and from the quarries create noise and dust emissions which can often be a source of

Figure 1.54 Visual intrusion of a quarry

annoyance to both residents and visitors. Transportation of the limestone causes particular problems, with heavy traffic concentrated on certain minor roads.

Tourist authorities have suggested restricting such traffic but hauliers are keen to increase the maximum gross weight allowed and thereby to allow more economic transport of quarry products.

Swinden quarry

Swinden Quarry (Figure 1.55) has been worked for limestone extraction and processing since 1900. In 1996 planning permission was granted to deepen the quarry by 100 m and to extend the working life of the operation up to the year 2020. Although planning permission had been refused for a previous application in 1993, the alternative proposal presented by the quarry owners Tilcon in 1996 addressed many of the concerns of both local residents and the planning authority:

'To meet the higher environmental standards desired in today's world, Tilcon is in the process of implementing environmental management systems throughout its operations with Swinden Quarry being one of the first'.

◆ **Landscape** The majority of the processing plant, which previously caused a severe visual impact,

Figure 1.55 Swinden Quarry

Figure 1.56 Swinden Quarry after restoration

has been moved to a new location 20 m below the existing quarry floor and entirely within the quarry (Figure 1.56). Mineral extraction will be confined within the existing quarry boundaries and the remaining landscape of Swinden Hill retained by quarrying to a depth of 100 m from beneath the remainder of the quarry floor.

◆ **Transport** Tilcon has recently upgraded rail facilities at the quarry and a new train has been introduced. It is estimated that this will result in a 23% decrease in heavy lorry movements – that is, 173 fewer HGV movements every day. Road haulage will still account for most movement of quarried material and the company insists on several environmental measures including the compulsory 'sheeting' of all vehicles – that is, the covering of all lorries carrying quarried stone with tarpaulins, so that dust and rocks do not escape.

◆ **Environment and community** The extension of the existing mineral waste disposal area will be progressively landscaped to reinforce the northern slopes of the hill. At the end of its working life the quarry will be restored and a nature reserve created, with a lake occupying the site. A liaison group has been created in association with the

local parish council to ensure that any concerns about the operation of the quarry can be addressed.

Recreation and leisure in the Yorkshire Dales

Almost ten million people live within 90 minutes' driving time of the Yorkshire Dales National Park (Figure 1.57). It is estimated that over nine million day visitors use the park every year, with the

Figure 1.57 Sources of visitors to the Yorkshire Dales National Park

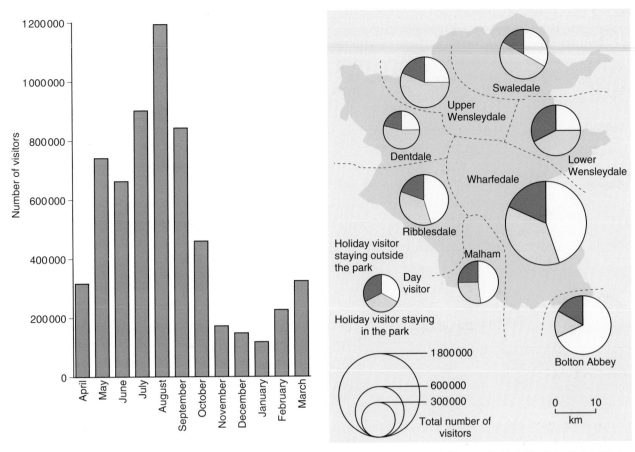

Figure 1.58 Monthly visitors to the Yorkshire Dales National Park

Figure 1.59 Distribution of visitors to the Yorkshire Dales National Park

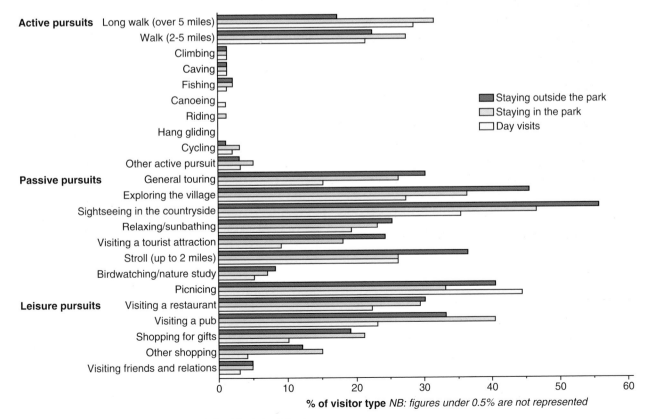

Figure 1.60 Activities in the Yorkshire Dales National Park, by visitor type

Table 1.2 Tourism-generated jobs in the Dales

Job type	Numbers of jobs
Food shops	50
Specialist shops	239
Cafes/pubs	446
Attractions	32
Garages	80
Public transport	2
Parking	37
Serviced accommodation	400
Self-catering	452
Hostels/bars	60

Source: Yorkshire Dales Visitor Study, 1991

numbers reaching a peak during the summer months (Figure 1.58). In the southern part of the park, day visitors are particularly important, reflecting the proximity and ease of access from the conurbations of West Yorkshire, Greater Manchester and Merseyside (Figure 1.59). The local economy has benefited substantially as visitors create increased employment and income, especially since a wide range of activities are spread across the park (Figure 1.60). A visitor study carried out during the 1990s suggested that spending by park visitors amounted to over £89 million annually and figures suggest that the tourist industry supported the equivalent of 1000 full-time jobs in the park and perhaps another 250 outside (Table 1.2). The money also has a knock-on effect, generating income from materials and services which are bought by local tourist businesses. The increasing number of visitors is now placing a considerable strain on certain popular areas within the park and is causing environmental problems which include:

◆ severe footpath erosion

◆ disturbance to farm livestock and wildlife

◆ traffic congestion on certain routes at peak times

◆ parking problems at visitor attractions, leading to damage to village greens

◆ increased commercialisation of traditional villages

◆ pressure for new developments to provide for and to attract visitors.

Some areas within the park attract a lot of people and these **honeypot** sites need to be sensitively managed. One such site is the Malham area (Figure 1.61). Few other places in the UK possess upland limestone scenery on such an impressive scale. The importance of the area has recently been recognised by English Nature which has designated it an SSSI and the area around Malham Tarn as a **National Nature Reserve (NNR)**.

This area has been popular with visitors for over 200 years, and although Malham village had a population of only 134 at the last census, the huge numbers of visitors have helped to maintain local services, including hotels, cafés, shops and pubs which might otherwise have closed. At certain times of the year the levels of traffic and car parking can substantially reduce the appeal of the village (Figure 1.62). Official car parks quickly fill up and vehicles park on roadsides and in passing places on the narrow roads. An experimental traffic management scheme during the early 1990s was abandoned due to pressure from local people and traders who reported loss of income during the period of the experiment. There are no plans to extend car parking in the village because the existing car park is often empty at weekdays in winter. During the summer peaks, a local landowner provides additional car parking space.

Figure 1.62 Malham village

Malham Cove
This great limestone amphitheatre was formed when ice and water eroded the hillside back from the line of the Mid-Craven Fault. At times it must have been a spectacular waterfall, as abundant meltwater flowing down the Watlowes valley in the ice age tumbled 70 m into the ice-filled Cove.
Obviously, with no surface flow across the limestone plateau today, the Cove remains dry, except in times of severe flooding (such as 1969). The Cove face is one of the few natural nest sites for housemartins, and is a refuge for rare plants like the Pennine whitebeam.

Malham Tarn
This attractive lake lies on a bedrock of slates in a depression scoured out by glacier ice in the ice age. The occurence of Silurian slate, surrounded on all sides by younger Carboniferous Limestone, is the result of earth movements and erosion in the geological past. The Tarn is scientifically important because springs, rising from the surrounding hills, bring in dissolved limestone, making it the highest lime-rich lake in the country. The unusual habitat of the Tarn is therefore a home for a unique collection of plants and animals.

Gordale Scar
Though on the same scale as Malham Cove, Gordale Scar is dramatically different. It is a twisting gorge, eroded through the thickly bedded limestone by meltwater streams in the ice age. The modern stream is much smaller than former ones but is notable because it is rich in dissolved limestone. This precipitates through-out the length of the gorge to form large deposits of yellowish tufa. Another tufa deposit may be observed further downstream at Janet's Foss.

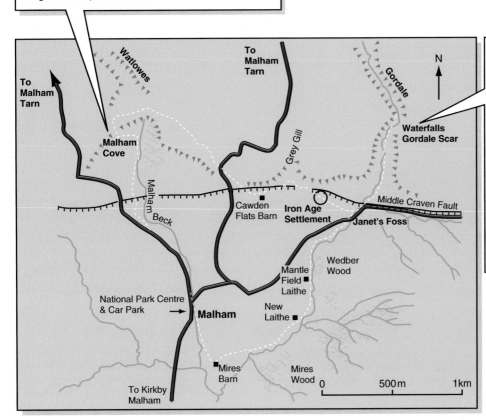

Figure 1.61 The Malham area

Many visitors use the local services but they also put pressure on landscape resources. Visitors walking to Malham Tarn and the Cove cause footpath erosion and twenty years ago it was necessary to surface the main path to the Cove to combat the excessive erosion. Since then most of the other popular paths in the area have also been surfaced.

Even the limestone pavements have been subject to degradation by removal of stone, and wear from visitors' feet. The pavements are now protected by local bye-laws, and regularly monitored by wardens. Although litter is a constant problem and a risk to livestock and wildlife, there is a policy not to provide litter bins in the area and to encourage visitors to remove their own litter. The litter problem has been substantially reduced but it is still necessary to organise litter picks in the most popular visitor areas.

QUESTIONS

6 a What are the main problems associated with limestone quarrying in the National Park?
b How has Tilcon addressed these issues at Swinden Quarry?

7 Describe and explain:
a the monthly distribution (Figure 1.58) and
b the spatial distribution (Figure 1.59) of visitors to the Yorkshire Dales.

8 a What particular problems does the management of honeypot sites such as Malham pose for planners?
b What efforts have been made to overcome these problems?

A great deal of publicity has been given in recent years to the concept of **global warming**. It is widely thought that the world is currently entering a warm period, with average temperatures increasing and sea levels rising due to the melting of ice caps. Such a change is part of a cyclical process. Research has revealed that if solar radiation is reduced over a long period of time (causing lower average temperatures) glaciation will take place.

During the past 2.5 million years there have been periods of intense cold when snow accumulated in

Figure 1.64 Glacial landscape features

the mountains of Scotland, northern England, Wales and Ireland and formed **glaciers**. During extended periods of cold climate, these glaciers joined up to form vast ice sheets. In between these cold periods, during **interglacials**, the glaciers began to disperse from these centres of ice accumulation (Figure 1.63). As they moved outwards their effect on the landscape below the ice was dramatic and the Scottish Highlands, the Lake District and Snowdonia were all etched with the characteristic features of glaciated upland landscapes (Figure 1.64).

Glaciation in the UK

During the ice ages, the extreme southern part of the UK, south of the River Thames and the Bristol Channel, was not glaciated (Figure 1.63). This region experienced a tundra-like climate, similar to areas such as northern Canada and Alaska today. For most of the year the ground was permanently frozen **permafrost**. These areas were subject to seasonal surface melting of the frozen ground, and the resulting water was swollen by the meltwaters draining from the ice sheets further north. In chalk areas such as the North and South Downs, **fluvioglacial erosion** by this meltwater is thought to be responsible for the formation of the characteristic dry chalk valleys.

In contrast, most of the eroded sands and gravels from the meltwater were washed into the valleys of the main rivers such as the Severn and the Thames,

A	North-west Highlands	F	Snowdonia
B	Cairngorm Mountains	G	Central Wales
C	Southern Uplands	H	Northern Ireland
D	Lake District	I	South-west Ireland
E	Southern Pennines		

Figure 1.63 Ice movement and centres of ice dispersal

where they were later carved into **river terraces** above their flood plains by the action of the rivers.

Glacial landscapes and processes in Scotland

Of all geomorphological systems, glaciation has had the greatest impact on Scottish scenery. Figure 1.65 shows that glaciation has been very selective in its effects, with a marked contrast evident between the eastern and western Highlands. The west consists of a rough ice-scoured landscape where the mountains have been heavily dissected by glacial troughs and valleys. The east shows much less landscape modification with evidence of erosion often confined to the valleys.

Scale: 0 — 100 km

Legend:
- Mountains & Plateaux heavily dissected by troughs & corries
- Areal scouring
- Possible light areal scouring
- Streamlining in rock & drift
- Streamlined drifts (drumlins)
- Selective linear erosion
- Intermontane basins
- High rolling terrain with a largely fluvial valley pattern
- High rolling terrain deeply dissected by fluvial valleys

Figure 1.65 Glacial landscapes in Scotland

To some extent, these variations can be explained by the different glacial histories of the east and west of the country. The west had a maritime climate and glaciers built up first in these environments and remained there for longer. In addition, there were periods such as the **Loch Lomond Re-advance**, 11 000 years ago, when glaciers only existed in the west. It has been suggested that the greater evidence of glaciation in the west is therefore mainly due to longer exposure to glacial activity.

Glaciers may either be frozen to the underlying rocks or be able to slide over them. Erosion takes place when glaciers slide, but this will only happen if the temperature at the base of the glacier is close to melting point. When the ice base temperature is well below 0 °C, the lowest ice layer is frozen onto the underlying surface and therefore much less erosion takes place. During the ice ages, the western part of Scotland was warmer and wetter, and the underlying slopes were steeper than in the east of the country. Together, these factors favoured fast-moving ice and conditions ideal for erosion. The most intense erosion took place in the west where the lochs and valleys are deepest, sometimes forming **fiords** or heavily glaciated coastal valleys. In the east, conditions were drier and colder and the slopes more gentle, resulting in much less ice movement and therefore less erosion. Due to the varied preglacial landscape, particularly the bedrock and channel system, the local effects of glaciation varied considerably, and different parts of Scotland now show different types of landforms and different intensities of erosion. These factors affecting the rate of glacial erosion are summarised in Figure 1.66.

QUESTIONS

1 Why were the upland areas of the UK the main centres for ice dispersal after the ice age?
2 Using Figures 1.65 and 1.66, describe and explain the differences in glaciated landscapes in the east and the west of the Scottish Highlands.

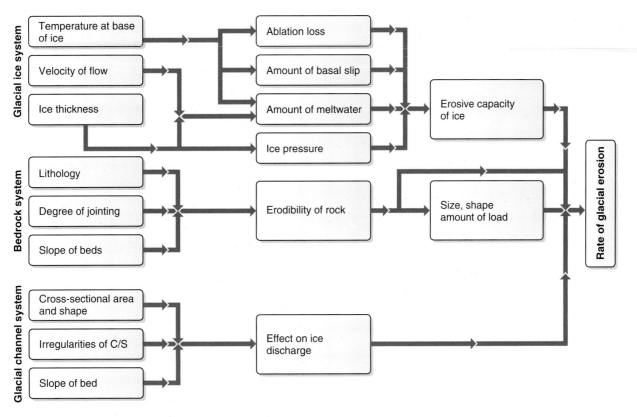

Figure 1.66 Factors affecting rates of glacial erosion

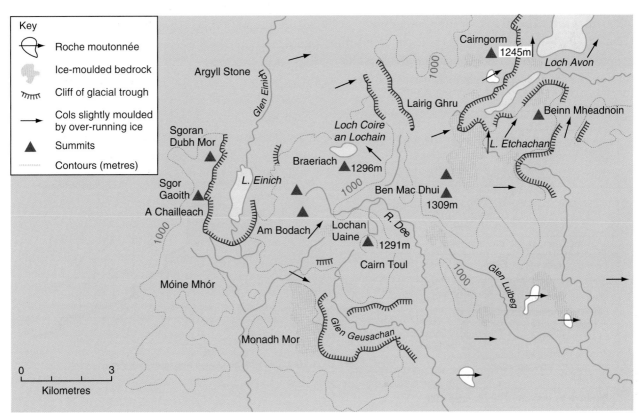

Figure 1.67 Glacial erosion in the Cairngorms

Figure 1.68 The Cairngorm plateau

The Cairngorm mountains

On a local scale, the influence of geology and pre-existing landforms on ice flow during glaciation was responsible for the patterns of glacial erosion. The plateau of the Cairngorm mountains, situated between the valleys of the rivers Dee and Spey in northern Scotland, forms the largest area of land over 1000 m above sea level anywhere in the UK. Within the Cairngorms (Figure 1.67) there are four summits over 1200 m: Ben Mac Dhui (1309 m), Braeriach (1296 m), Cairn Toul (1291 m) and Cairngorm (1245 m). Between these peaks stretches an undulating granite plateau (Figure 1.68) containing classic glacial landforms such as deep troughs and massive **corries**

gouged out by glaciers during the last ice age (Figure 1.67). The ice direction, moving from west to east, can be identified from the orientation of **roches moutonnées**, streamlined glacial features aligned in the direction of the former ice flow (Figure 1.69).

The plateau surface

The surface of the Cairngorm plateau is not completely flat but comprises smooth rolling slopes and shallow valleys, contrasting sharply with the deep glacial troughs cut by glaciers. The Cairngorms owe their rounded appearance to their granite composition which has weathered down to **scree** on

Figure 1.69 Formation of a roche moutonnée

Figure 1.70 The Cairngorm tors

the plateau tops by the enlargement of joints in the rock by freeze–thaw action.

Cairngorm granite has a high feldspar content and is pink in colour. It was formed from molten lavas pouring out of the Earth's crust 400 million years ago. In addition to glaciated landforms, the Cairngorms also exhibit features developed by the erosion of granite.

The Cairngorm **tors** are the finest in Scotland and consist of gigantic blocks of rocks (Figure 1.70) rising up to 25 m above the smooth hill slopes. Tors were formed before the ice ages when the climate was hot and wet. The granite was deeply weathered to an

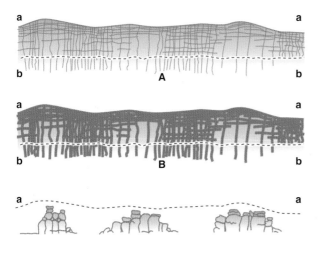

Diagram to illustrate the formation of tors. An original rock surface with variations in joint frequency (A) is attacked by chemical weathering to form areas of deeply rotted rock and irregular remnants of fresh rock (B). In the bottom drawing the deeply rotted rock has been stripped by erosive processes to leave the fresh rock tors. (After Linton 1955)

Figure 1.71 The formation of tors

uneven depth, leaving areas of harder bedrock between areas of more heavily weathered and crumbling granite. During the ice ages the rotted granite was eroded, leaving behind some of the larger blocks (Figure 1.71). Many have since weathered into fantastic shapes influenced by rock jointing. Tors are particularly common on the eastern Cairngorm summits such as Ben Avon and the Barns of Bynack.

The plateau edges and the valley

On the edges of the Cairngorms, small mountain glaciers have cut other landforms. During the ice ages snow accumulated in armchair-shaped corries consisting of rock basins overlooked by rugged cliffs (Figure 1.72). Today these basins are often deep enough to hold small **lochans** such as Coire an Lochain to the north of Braeriach and Lochan Uaine north of Cairn Toul. An analysis of the distribution of the Cairngorm corries reveals that most face north and east, because the glaciers built up more quickly on these shady slopes where the accumulating snow was more protected from the sun's rays. The accumulated snow hardened to form ice.

Subsequent erosion through **nivation** (snowpatch erosion), freeze–thaw action, **solifluction** (soil flow), meltwater erosion and chemical processes, deepened the corries. Eventually the accumulated ice began to flow away from the steep back wall of the corrie by **rotational sliding**, further deepening the corrie as it did so (Figure 1.73). Where several corries occurred close together, they were the source for the larger valley glaciers (Figure 1.74) which excavated deep **U-shaped glacial troughs** such as the great cliffed valleys now occupied by Lochs Avon and Einich (Figure 1.75). Some valleys, such as the Lairig Ghru, cut through major watersheds and became important routeways through the mountains (Figure 1.76).

When the ice began to melt, erosion became intense and created a number of fluvioglacial landforms resulting from meltwater and massive deposition of glacial debris. Much of the meltwater flowed through Strathspey where the river volume increased, particularly during the summer melt periods. The swollen rivers Feshie and Spey deposited large quantities of boulders, sand and gravel on the valley floor which have subsequently been cut into by the

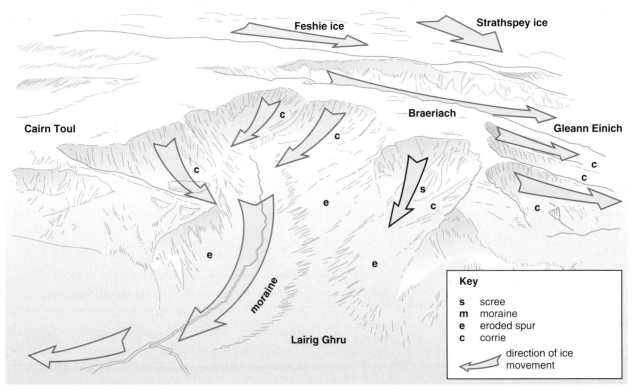

Figure 1.72 Corries of Braeriach and Cairn Toul

Randkluft, between ice and back wall

Bergschrund opens up as ice pulls downhill

Steep back wall

Extending flow

Extending flow

Compressive flow

Ice fall

Rotational sliding

Ice

Bedrock

Deepening caused by abrasion

Rock barrier or threshold

Figure 1.73 The development of a corrie

Figure 1.76 The Lairig Ghru

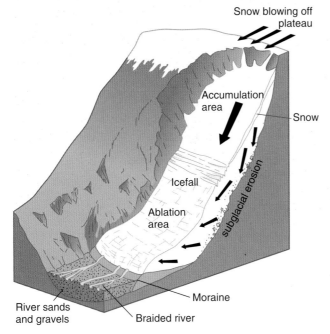

Snow blowing off plateau

Accumulation area

Snow

Icefall

subglacial erosion

Ablation area

Moraine

River sands and gravels

Braided river

Figure 1.74 The formation of a valley glacier

rivers, creating terracing and **braided river channels** where the water flows through several channels (Figure 1.77).

The Cairngorms exhibit a landscape typical of selective **linear erosion**. There was minimal erosion at the base of the ice sheet which was frozen to the bedrock for much of the time, making it relatively protective over the high ground of the Cairngorm plateau. In contrast, where the ice sheet was channelled into valleys, it was thicker, warm at the base and therefore erosive. This helps to explain the differing glacial landscapes within the

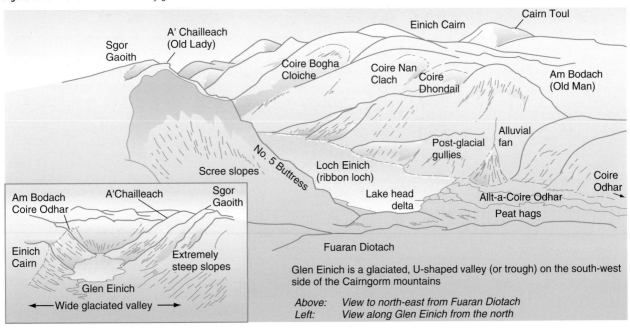

Cairn Toul

Einich Cairn

Sgor Gaoith

A' Chailleach (Old Lady)

Coire Bogha Cloiche

Coire Nan Clach

Coire Dhondail

Am Bodach (Old Man)

No. 5 Buttress

Scree slopes

Loch Einich (ribbon loch)

Lake head delta

Post-glacial gullies

Alluvial fan

Allt-a-Coire Odhar

Coire Odhar

Peat hags

Am Bodach Coire Odhar

A'Chailleach

Sgor Gaoith

Einich Cairn

Extremely steep slopes

Glen Einich

Wide glaciated valley

Fuaran Diotach

Glen Einich is a glaciated, U-shaped valley (or trough) on the south-west side of the Cairngorm mountains

Above: *View to north-east from Fuaran Diotach*
Left: *View along Glen Einich from the north*

Figure 1.75 Glen Einich – a glaciated valley

Figure 1.77 River Feshie

Cairngorms where deep glacial troughs such as that occupied by Loch Avon contrast with the plateau surface, dotted with tors and weathered rocks – features that would not have survived beneath an active ice sheet.

QUESTIONS

3 Using diagrams, explain the formation of both roches moutonnées and corries. For each feature, explain the significance of their spatial pattern in the Cairngorms.
4 Describe the processes involved in the formation of tors and explain their occurrence on the Cairngorm plateau.
5 Make a fully annotated copy of Figure 1.75, identifying and explaining the main features of the glaciated landscape shown on the sketch.

Conservation in the mountain areas of Scotland

A study of the National Parks in England and Wales (Figure 1.41) shows that many have been designated in areas of upland mountain scenery. England's first

Figure 1.78 Scotland's National Parks and National Scenic Areas

Footpath damage

Caravans
Wild caravanning,
visual intrusion

Ski areas
Need for better
design, management
and maintenance

Intrusive activities
Jet skiing, off road
cycling

Ecological damage
Loss of heather
moor due to
deer/grouse
management

Figure 1.79 Tourism and the Scottish environment

National Park was created in the 1950s but surprisingly, although Scotland has had many parts designated as National Scenic Areas (Figure 1.78), it did not have any National Parks until 2002.

During the 1990s it became increasingly clear that certain parts of Scotland were coming under heightened visitor pressure largely due to three main factors:

◆ increased affluence of the population

◆ greater leisure time

◆ improvements in personal mobility (cars, communication links etc.).

Figure 1.79 indicates just some of the problems and pressures which tend to be concentrated at certain popular scenic locations.

In February 1999 **Scottish Natural Heritage (SNH)** published advice to the Scottish Executive on how National Parks in Scotland could operate and, in 2000, primary legislation was approved by the Scottish Parliament for the establishment of the first two National Parks in Scotland. Scotland's first National Park, the 1865 km² Loch Lomond and the Trossachs National Park, opened in July 2002. The

Cairngorm National Park became Scotland's second National Park in September 2003. Future National Park locations may include Wester Ross and Glencoe (Figure 1.78).

Designation of these areas as National Parks has allowed improved integrated and effective planning and management of these areas. However, it is important to remember that the creation of National Parks is a double-edged sword; they do provide greater environmental protection but at the same time they heighten the profile of these areas which will encourage yet more visitors, thereby intensifying the existing pressures on the landscape.

QUESTIONS

6 What evidence is there that certain areas of Scotland are in urgent need of stricter visitor management policies?

7 One of the landowners in Wester Ross who opposes the creation of a National Park in that area has argued that it 'would create a honeypot effect'. Explain what you think he means by this and outline whether or not you agree.

Conservation issues in the Cairngorms

The Cairngorms National Park (Figure 1.80) is home to 17 000 people, living in substantial towns, villages, hamlets and houses in the countryside. At 11.6 people per square kilometre, the population density is very low. The major centres of population are Aviemore, Ballater, Braemar, Grantown-on-Spey, Kingussie, Newtonmore, and Tomintoul. The park covers a very large area, 3800 km², which is 40% larger than the Lake District and twice the size of the Loch Lomond and the Trossachs National Park. It is a mountainous area, with four of Scotland's five highest mountains within the park boundaries. It contains 52 mountain summits over 900 m, 10% of the total land area is over 800 m in height and 68% is over 400 m above sea level.

This high altitude, combined with the northern latitude of the Cairngorms, has inevitably had a marked influence on the local climate, which is the most continental in the UK. Braemar, on the eastern slopes, is regularly the coldest place in the country

Figure 1.80 Cairngorm National Park

Figure 1.81 Cairngorm weather station in winter

during winter months. There is a weather station on the summit of Cairngorm (Figure 1.81) which, on average, records 196 days each year with air frost, and 83 days with temperatures below zero. Due to these severe climatic conditions, snow lies longer on these mountains than in any other part of the UK, usually for about 90 days at the summit of Coire Cas. This severe climate and the thin, infertile soils which have been weathered from the granite have

combined to prevent even extensive farming outside the lower and more sheltered glens. As a result the Cairngorms are one of Europe's last remaining **wilderness** areas.

An area equivalent to 39% of the park area is designated as important for nature heritage; 25% is of European importance. Flora and fauna are particularly rich in the Cairngorms, and for this reason it became the UK's first National Nature Reserve in 1954. This was an attempt to preserve the unique arctic and alpine plant associations such as moss campions and lichens, and the varied wildlife which includes birds such as the osprey, golden eagle, ptarmigan and snow bunting. Indeed this National Park is home to 25% of the UK's threatened bird, animal and plant species.

The Cairngorms has an important and unique range of natural heritage resources in a local, national and international context and consequently deserves special attention. An area of 147 194 hectares, or 39%, of the Cairngorms comprises national and international conservation designations, such as Sites of Special

Scientific Interest (SSSIs), Special Protections Areas (SPAs), candidate Special Areas of Conservation (cSAC), Ramsar sites, National Nature Reserves (NNRs) and National Scenic Areas (NSAs). The Cairngorms is also a candidate World Heritage Site (see page 18) for Earth Heritage, containing as it does the finest collection of glacial landforms in the UK.

Although it cannot be regarded as a wilderness area when compared, for example, with the Canadian Arctic, it is nevertheless remote by most European standards, and is the largest tract of land distant from public roads in the UK. This remoteness, together with the superb opportunities for winter recreation (due to the best snow-holding conditions in the UK), summer rock climbing, hill-walking and birdwatching, have resulted in a large increase in the number of people visiting the Cairngorms. Visitors come here all year round and from an increasingly urban population. With better accessibility via the improved A9 road, pressures on the area have increased rapidly and inevitably conflicts have developed between those keen to expand facilities for tourism, and conservation interests. It is thought that at least 500 000 people visited the Cairngorms in 2001 – 350 000 to Badenoch and Strathspey alone. Tourism-related businesses account for about 80% of the economy, including activities such as skiing, walking, fishing, shooting and stalking.

The expansion of skiing in the Cairngorms

One of the fundamental contradictions of modern countryside recreation is that as more people are encouraged to seek out remote areas, their increased numbers make a greater impact on these regions, thereby making them less attractive as wilderness areas. In order to solve this problem the concept of **sustainability** (meeting today's needs without reducing the ability to meet tomorrow's needs) has become an important consideration in many development projects.

Since the 1960s walkers, climbers and wildlife enthusiasts in the Cairngorms have been joined by increasing numbers of skiers. Whereas the former groups make minimal demands on the landscape, the popularity of the Cairngorms for skiing has been accompanied by growing demands for better facilities

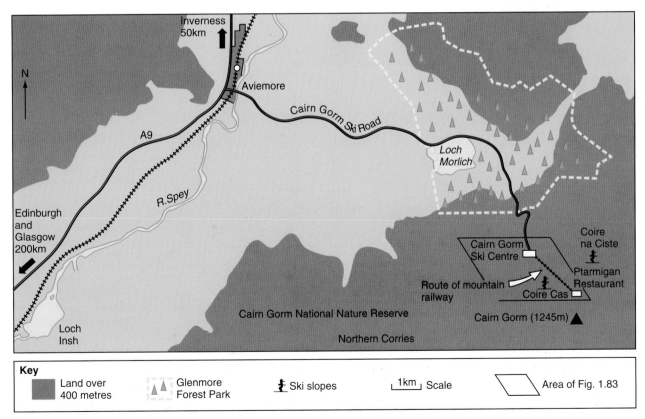

Figure 1.82 Aviemore and the Cairngorms

such as new access roads, more ski runs and ski tours, restaurants and accommodation.

The expansion of skiing in the Cairngorms has played a vital role in the post-war local economy of the Spey Valley. Throughout the Scottish Highlands, lack of employment opportunities has frequently led to **rural depopulation**, particularly amongst the younger age groups. Yet in the Spey Valley, tourism-related employment (40% of employees work in businesses directly related to tourism) has meant that jobs are available locally. A recent estimate suggested that skiers bring in about £12 million annually to the local economy, generating the equivalent of 350 full-time jobs.

Mechanised downhill skiing in the Cairngorms began in 1961 with the construction of the White Lady chairlift and the extension of an access road through Glen More from Aviemore (Figure 1.82). Downhill skiing was concentrated at Coire Cas and Coire na

Ciste. By 1980 these areas had been developed to their capacity, and by the mid-1980s the Cairngorm Chairlift Company produced plans for further expansion of skiing into an area known as the Northern Corries, which had previously been used principally for climbing, walking and cross-country skiing. The planning application was refused in 1990, largely after **Scottish Natural Heritage (SNH)**, the countryside advisory agency for Scotland, advised the government against the development on the grounds of visual impact, the effect on other forms of recreation and the intrusion of commercial development into an area of high and wild land.

But the existing 1960s ski facilities could not meet the increasing demands for skiing each winter. The opening of new modern facilities elsewhere in Scotland, notably at Aonach Mor in the Nevis Range near Fort William, and also the attractions of European ski resorts (combined with low-budget

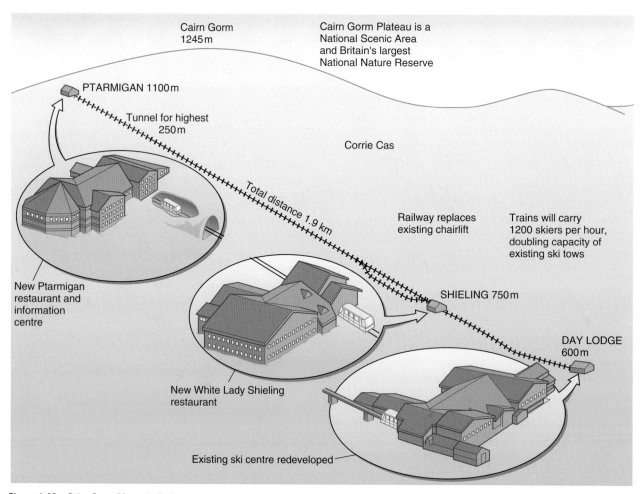

Figure 1.83a Cairn Gorm Mountain Railway

Figure 1.83b

European airlines) have had a serious economic impact on skiing in the Cairngorms. The facilities in the Cairngorms were older and unreliable and the chairlift had to be stopped whenever the winds exceeded 30 mph (and this in one of the UK's windiest areas). In 1997 the government gave its backing to a new scheme to replace the chairlifts with a funicular railway (Figure 1.83a and b). The developers argued that the railway would provide a year-round tourist attraction and replace much of the existing piecemeal development with a single scheme.

Construction work on the project began in August 1999 with huge opposition arising primarily from environmentalists. In a letter to *The Times* newspaper, mountaineer Chris Bonington argued that the development 'would do irreversible damage to one of Europe's most important wild areas', and an article in *The Scotsman* newspaper went further and commented:

The Cairn Gorm funicular railway is a nonsense, a foolish scheme to rescue the ski resort from the threat of snowless winters by allowing summer visitors to damage the resource that brings people in the first place. And what a place it is, a world of sky and space, of rock and water, forest and loch. A place we should be proud of, a place we should defend and protect rather than fill with tons of concrete and install a toy train.

Developers argued that such schemes have been in operation in the Alps for many years, operating without the kind of environmental Armageddon predicted by the Cairngorm conservationists.

Throughout the 1980s and 1990s, SNH were the principal objectors to any development plan for the expansion of skiing in the Cairngorms. A precondition for withdrawing their objections to the funicular plan was the acceptance of a strict 'Visitor Management Plan'.

The funicular railway was finished in 2001. The 'Visitor Management Plan' prevents users of the railway from leaving the top station, unless they are going to ski down the mountain during the winter season. Visitors are allowed to visit a café, interpretation centre and shop, but they will not be allowed to climb to the summit of Cairngorm (as the chairlift clients were able to do). Those who wish to visit the wild areas on the summits are only able to do so via the long walk in from the ski centre car park (that is, they must walk up the mountain). This has caused upset to many climbers and ski mountaineers who claim they cause minimum impact on the landscape (compared with downhill skiing) but were previously allowed to use the chairlift to gain access to the Northern Corries and the remote Loch Avon basin.

The debate over the railway is a classic example of how polarised views between developers and conservationists have greatly hampered conflict resolution. As one observer commented: 'A fence was built, and everyone was forced to choose on which side they wanted to stand.' It may be that in future, dialogue and negotiation should be the basis of planning disputes such as this, the outcome of which have such profound socio-economic and environmental consequences.

QUESTIONS

8 What physical factors are responsible for the Cairngorms being one of Europe's last wilderness areas?

9 Why did the Cairngorms become a National Nature Reserve in 1954 and a National Park in 2003? What protection does this status provide?

10 Why are people in the UK increasingly attracted to remote areas for recreation?

11 Outline the arguments for and against the building of the Cairn Gorm Mountain Railway.

The changing village: Laggan and the conservation of people

The village of Laggan is situated in the upper part of the Spey Valley (Figure 1.84) at the point where the river is bridged by the A86 road which links the area with the west coast of Scotland. Two hundred years ago records indicate that the population of the parish was over 2000 people; the 1991 census put the population at 189.

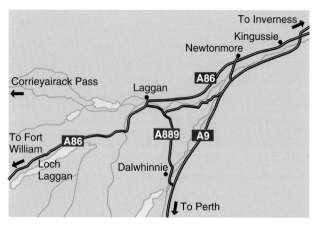

Figure 1.84 The location of Laggan

Table 1.3 Employment in Laggan, 1970 and 1990

Laggan employment	1970	1990
Farmers	6	6
Farm employees	19	11
Estate workers	13	7
Crofters	5	4
Forestry workers	14	3
Conservationists	0	2
Arts & crafts	0	4
Retired locals	25	11
Retired incomers	0	11
Self-employed foresters & fencers	0	6
Others living in Laggan, employed outwith Laggan	0	4

The depopulation of Laggan began during the nineteenth century with the **clearances** as landowners evicted crofters and small tenant farmers and used the land to increase their own sheep flocks. Thousands of Highland Scots were forced to leave the area; some migrated to take jobs in the expanding industries of the cities of central Scotland while others emigrated to Canada and the USA.

Since then depopulation has continued; with increased mechanisation there are fewer jobs available on the remaining farms and estates (Table 1.3). Apart from the lack of employment, another problem for local people is the lack of affordable housing. The growth of the holiday homes market, coupled with the arrival of 'new settlers' in the form of conservationists, artists, craftsmen and retired people from the south, have all served to increase house prices and limit the amount of property available to people living and working in the area.

In Laggan the local community raised enough funds through government, European Union and development agency grants to build a community hall and television reception has been greatly improved by the local people erecting their own mast.

Changes in the village in the late twentieth century are summarised in Table 1.4. The decline in pupil numbers at the village school was a matter for concern in the 1990s, but the situation has been reversed with over half of the current pupils being from English families who have moved into the area. Tourism is clearly the growth area, helped by a recent popular BBC television series, 'Monarch of the Glen', which was set in the area (Figure 1.86). This show was sold to 12 countries and an estimated 50 million people worldwide are expected to see the programme. This has had the effect of attracting more tourists and visitors to Laggan, as well as bringing income directly to the village during the filming of the programmes. The village has retained its general store (Figure 1.85). One recent survey found that as many as 9% of tourists choose to visit the Highlands because they have seen the scenery in films and television programmes. In June 2001 the local tourist board decided to capitalise on the success

Table 1.4 Changes in Laggan, 1990–2000

1970	1990	2000 changes
School 35 pupils, 2 teachers	**School** 16 pupils, 1 teacher, 1 part-time teacher	**School** 29 pupils (5 in nursery), 2 teachers, 1 assistant, 1 auxiliary, peripatetics
Church	Church	
Shop & petrol pumps (privately owned)	Shop & petrol pumps	
District nurse	District nurse (based Kingussie)	
Doctor	Doctor & Associate	
Village Hall (needing repair)	New Community Hall	
1 pub	2 pubs	
1 hotel	2 hotels	Bed & Breakfast and Guest Houses
4 self-catering cottages	8 self-catering cottages	
2 holiday houses	4 holiday houses	1 caravan site
11 forestry houses	Forestry houses 3 privately owned, 5 rented, 3 empty	All 11 now in use
2 empty cottages	Cottages now bothies	Bunkhouse for climbers
	5 new houses	

Figure 1.85 Laggan general stores

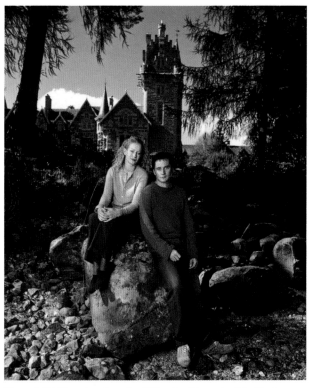

Figure 1.86 Characters from the BBC series 'Monarch of the Glen'

of the programme by renaming the area 'Monarch Country', a title which will be used on brochures and road signs. It is estimated that the area could benefit by £100 million in increased tourism revenue by 2010. The dependence on tourism is therefore increasing, while traditional occupations continue to decline.

Forestry in the Cairngorms

Although forests in Scotland now cover 1.4 million hectares or 15% of the land area, this is still well below the extent of woodland cover of most countries in mainland Europe. In Scotland, the expansion of forestry during the late twentieth century was heavily criticised for the poor appearance of new planting, the intensive management techniques used, and the use of financial and tax incentives which encouraged low-yield planting on poor-quality land, to the detriment of the natural environment.

The government's forestry policy is now based on multiple objectives: wood production, **biodiversity**, **landscape diversity** and recreational provision. Forestry supports over 10 000 jobs in Scotland, the majority of them in rural areas. Despite the use of mechanised techniques, the increase in timber production from 3 million cubic metres in 1994 to 8 million in 2014 will provide opportunities for further employment in wood harvesting and processing. A major problem now in many rural areas (notably Dumfries and Galloway), as timber harvesting increases, is the poor condition of many minor roads and weak bridges which will have to carry the weight of timber lorries.

The concern for maintaining biodiversity and landscape diversity is partly a response to public concern about the huge swathes of monotonous coniferous plantations that appeared throughout the Scottish uplands during the post-war period, and also a recognition of the importance of sustaining and expanding native woodland habitats. One forestry company, Highland Birchwoods, manages over 2000 hectares of native birch woodland. Recent examples of their successful projects include:

◆ the commercial cultivation of culinary shiitake mushrooms on small roundwood which would otherwise have had little value

◆ the use of new manufacturing techniques to produce flooring from timber which would otherwise have been used for firewood.

The importance of the woodland environment in providing opportunities for a wide range of recreational pursuits has long been recognised by the five Scottish Forest Parks, one of which is at Glenmore in the Cairngorms (Figure 1.82). The attraction of the Cairngorms for visitors is as much based on the fringe of varied woodland on the lower slopes of the hills as it is on the high tops. In terms of visitor numbers, with the exception of downhill skiing, far more people enjoy their recreation in the woodlands than on the hilltops. Glenmore Forest Park provides space for many activities (for example, orienteering) that largely depend on the woodland environment. Development pressures are less than on the hilltops because the woodland is able to absorb large numbers of people and their cars in sheltered and secluded surroundings, and so they have minimum impact on the natural heritage. In addition, the woodland is a rich reservoir for wildlife; it also provides a bad-weather alternative to the hills. Forest Enterprise has recently embarked on a strategy of enhancing the forest for amenity, conservation and recreation purposes in an attempt to re-establish a native pinewood. Other schemes include the introduction of the **Community Woodland Supplement** to the Woodland Grant Scheme. By these means the government encourages new woodlands which local communities can use for recreation. Over 1500 hectares of community woodland have been planted in Scotland close to towns and villages since the scheme began in 1992.

QUESTIONS

12 Identify the similarities and differences in the problems facing Highland villages such as Laggan, and those in the Yorkshire Dales (pages 34–37).

13 Why is UK forestry policy now based on multiple objectives, only one of which is timber production?

14 Referring to specific examples, identify successful attempts at forest diversification.

The Highlands and hydro-electricity

The high precipitation rates and the natural lochs of the Scottish Highlands make it an ideal place to develop renewable energy schemes, in particular hydro-electric power.

Many people are familiar with the large-scale 1950s hydro-electric power schemes that characterise many of the glens and lochs of Scotland (Figure 1.87).

Figure 1.87 Laggan HEP dam

Large hydro-electric power schemes cause **visual pollution** and can be responsible for severe **environmental impact**. As a result of this, most future hydro schemes in Scotland will be relatively small-scale schemes which do not involve the construction of a large dam or the creation of a new reservoir.

These 'new style' schemes are known as 'run-of-river schemes'. They generally involve the building of an intake weir to take water from the top of the river, and a pipeline carrying the water to a new powerhouse (Figure 1.88). They allow a basic amount of water down the river at all times to ensure its ecological health, but use the remainder for power generation. More than the basic amount is only released when the river is in spate (flowing at bank full) and the surplus flows over the holding weir. Small upland rivers with a steep gradient are most suited to this type of development and the Highlands of Scotland has no shortage of such rivers.

There are currently twenty run-of-river schemes proposed for Scotland and a further five potential locations (Figure 1.89). One of the most recent schemes has been on the River Fechlin (Figure 1.90). Despite the need to improve the country's capacity to produce renewable energy, the building of such schemes has caused a lot of conflict between

A Run-of-River Hydro Scheme works by diverting the main flow of the river through a weir and into a separate tunnel/pipeline where energy can be generated. The four main parts of the scheme are:

1. The Diversion Weir
The weir is a concrete water trap designed to divert water. As it does not create a reservoir of any size or significance, flooding is kept minimal and the negative impact of the Project on both the environment and the local communities is limited. From here the water flows through a short tunnel into two large V-shaped underground chambers called 'desanders'

2. Desanding chambers
The desanders slow the flow of water and allow the collection and removal of any pebbles, sand and silt

3. The Pressure Tunnel
After the sediment is removed, the clear water is allowed to flow into the main tunnel

4. The Power Station
The steel pipe extends from the tunnel and into the power station where it directs the water through high-pressure jets, onto turbine blades. The turbines rotate like steel waterwheels at very high speeds and generate electricity

Figure 1.88 Run-of-river hydro schemes

Figure 1.89 Proposed new HEP schemes in the Highlands

rural landusers. In particular, kayakers and canoeists, who are the most frequent users of small upland rivers with a steep gradient, have seen the opportunities for their sport reduced. Whitewater canoeing, kayaking and rafting are some of the fastest growing sports in the UK and all are affected by run-of-river schemes. Over the years these local rivers have provided many recreational opportunities, boosted tourism and sustained local services in remote areas of Scotland, particularly the Highlands.

Global warming and renewable energy

Clearly global warming is an important issue and the burning of fossil fuels to create energy is a contributing factor. There are, however, many questions as to how effective small-scale run-of-river hydro-electric developments can be in contributing to a reduction in carbon dioxide emissions.

The remaining hydro-electric resource in Scotland is comparatively small. A report by the Scottish Executive entitled 'Scotland's Renewable Resource'

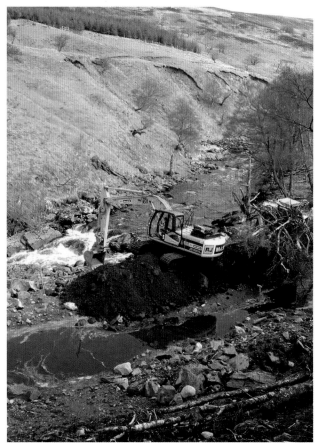

Figure 1.90 River Fechlin before and during construction of the run-of-river scheme

found that there was the potential to generate around 1000 GWH (Gigawatt hours) from small-scale hydro projects in Scotland. To put this in context, electricity demand in the UK is approximately 390 000 GWH. The report also identifies that much of this potential resource may be constrained by environmental and regulatory pressures. Offshore wind (45 000 GWH), wave (45 700 GWH) and tidal stream (33 500 GWH) offer a far more significant contribution.

The small capacity of many of the proposed schemes (Figure 1.89) could easily be achieved in other ways. For example, the 2 MW which can be produced by the Garrogie scheme on the River Fechlin can be compared to the output of a single wind turbine. Many environmentalists have argued that the Garrogie scheme was not worth the visual pollution and decrease in recreational possibilities for the amount of energy that the scheme creates.

Run-of-river hydro schemes also have a highly variable power output as they can only generate electricity after periods of rain. The energy cannot be stored and so they can only make a limited contribution to a steady supply of electricity.

The construction of hydro schemes is in itself very energy intensive and will involve significant creation of carbon dioxide emissions.

Hydro-electricity in Wester Ross

Wester Ross contains some of the finest coastal mountain scenery in the UK and is the home of a diverse array of wildlife including deer, pine martens and red squirrels. There is also a variety of marine wildlife and the possibility of spotting a seal, porpoise, dolphin or whale in the surrounding waters is very high. As a result the area is popular with tourists and tourism is the main source of income for the local economy.

In May 2003, Highland Light and Power proposed to build a small-scale hydro scheme in the Shieldaig/Slattadale area of Wester Ross (Figure 1.91). The proposed scheme was to involve four lochs in the Shieldaig and Slattadale forest area. If built it would have generated 3.55 MW of power, which is enough to power around 5000 households.

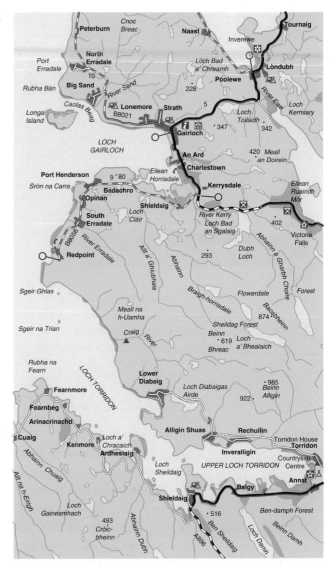

Figure 1.91 Shieldaig area

The Scottish Canoe Association, the Mountaineering Council of Scotland and Scottish Natural Heritage all objected to the proposed scheme and the project was rejected by Scottish Ministers in March 2004. Scottish Natural Heritage's main reasons for its objection were as follows:

◆ The area houses a large and thriving population of freshwater pearl mussels and the proposed project would be likely to have an adverse effect on them.

◆ The development would disturb over 1% of the (small) population of black-throated divers – an internationally protected species.

The development would lie within the Wester Ross **National Scenic Area (NSA)** and would have adverse impacts on the landscape and amenity of the

NSA. In the view of SNH this remote part of the NSA would qualify as 'wild land'. Although wild land is not designated in conservation terms, planning guidance cautions planning authorities to ensure that proposed developments do not compromise the wild land character of these areas.

The proposed run-of-river schemes are yet another example of how conflict can occur between different groups of people who all wish to exploit rural land resources.

The Beauly to Denny transmission line proposal

One of the main problems with potential alternative and renewable sources of energy in Scotland is that many of the production sites are in the north and west of the country, whilst the demand for electricity is highest in the central belt. Following eighteen months of public consultation held by Scottish

Hydro-Electric Transmission Limited (SHETL) and SP Transmission Ltd (SPL), the two companies have now published the proposed route for the 400 000 volt overhead transmission line which will replace the existing 132 000 volt line between Beauly and Denny (Figure 1.92). The main argument for the £320 million project is to allow new hydro-electric and wind power to reach the central belt of Scotland.

Environmental impact

Minimising the environmental impact of the replacement line along the route from Beauly to Denny has been one of the crucial factors for the companies involved. In keeping with this, visual effects have been assessed for properties and settlements, roads, railways, other rights of way, open spaces and recreation areas.

In order to reduce the visual impact, around 60% of the 220 km long line will be constructed on a route adjacent to the existing line, which it will replace. The existing transmission line between Beauly and Denny will be dismantled and removed once the new line becomes operational. The new line will be supported by around 600 pylons, which is approximately 200 fewer than the existing line, although they will be significantly higher.

Another key objective in selecting the proposed route has been to avoid proximity to as many residential properties as possible. There are currently around 90 residential properties located within 100 m either side of the existing transmission line. There will be 10 residential properties within 100 m of the proposed route for the replacement line.

As part of the final design phase, the location of individual pylons will be fine-tuned to reduce their effects on areas of particular sensitivity, for example, by using planting to minimise the effect on landscape views.

An alternative solution?

During the initial consultations, Highland Council, Cairngorms National Park Authority and Scottish Natural Heritage commissioned a study into the technical, financial and environmental issues relating to the feasibility of using underground transmission

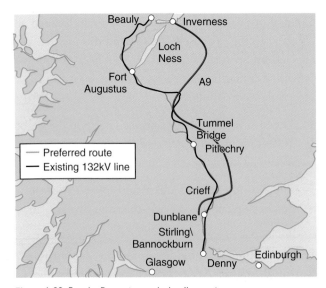

Figure 1.92 Beauly–Denny transmission line route

routes. However, this alternative to overhead power lines was dismissed for four main reasons:

◆ Underground power lines 'do not offer the same level of availability as overhead lines'. Prolonged faults are more frequent in underground cables than overhead lines and they also take, on average, longer to repair.

◆ Construction would involve a 'motorway-width disturbance' corridor and involve the movement of considerable volumes of spoil.

◆ There is potential to cause environmental damage from a fluid leak.

◆ An underground power line would cost between six and twelve times more than overhead lines.

QUESTIONS

17 Why do you think that much of the new power line will follow the existing power line route between Beauly and Denny?
18 Create a table outlining the visual, noise and other types of pollution that a project like this will create.

Key terms and concepts

◆ abrasion (p. 8)

◆ attrition (p. 8)

◆ backwash (p. 8)

◆ biodiversity (p. 60)

◆ braided river channel (p. 51)

◆ Carboniferous limestone (p. 27)

◆ clearances (p. 58)

◆ clint (p. 30)

◆ Community Woodland Supplement (p. 60)

◆ concordant coastlines (p. 10)

◆ constructive waves (p. 8)

◆ corrie (p. 48)

◆ corrosion (p. 13)

◆ dale (p. 27)

◆ deposition (p. 3)

◆ destructive waves (p. 8)

◆ discordant coastlines (p. 10)

◆ doline or shake hole (p. 30)

◆ dry valley (p. 29)

◆ English Nature (p. 31)

◆ environmental impact (p. 61)

◆ Environmentally Sensitive Area (ESA) (p. 34)

◆ erosion (p. 3)

◆ farm diversification (p. 33)

◆ fetch (p. 8)

◆ fiord (p. 10)

◆ fluvioglacial erosion (p. 45)

◆ footloose industry (p. 37)

◆ glacial drift (p. 30)

◆ glacier (p. 45)

◆ glaciokarst (p. 27)

◆ global warming (p. 45)

◆ gorge (p. 29)

◆ Great Scar Limestone (p. 29)

◆ gritstone (p. 39)

◆ gryke (p. 30)

◆ honeypot site (p. 42)

◆ hydraulic action (p. 8)

◆ igneous rocks (p. 2)

◆ interglacials (p. 30)

◆ karst (p. 27)

- landscape diversity (p. 60)
- limestone pavement (p. 30)
- limestone scar (p. 30)
- linear erosion (p. 51)
- lochan (p. 49)
- Loch Lomond Re-advance (p. 46)
- longshore drift (p. 9)
- metamorphic rocks (p. 2)
- National Nature Reserve (NNR) (p. 42)
- National Scenic Area (NSA) (p. 63)
- nivation (p. 49)
- permafrost (p. 30)
- permeable (p. 27)
- pothole (p. 30)
- river terrace (p. 46)
- roches moutonnée (p. 48)
- rotational sliding (p. 49)
- rural depopulation (p. 56)
- Scottish Natural Heritage (SNH) (p. 53)
- scree (p. 48)
- sedimentary rocks (p. 3)
- sink (p. 30)
- Site of Special Scientific Interest (SSSI) (p. 31)
- solifluction (p. 49)
- solution (p. 8)
- stalactite (p. 30)
- stalagmite (p. 30)
- sustainability (p. 55)
- swallow hole (p. 30)
- swash (p. 8)
- tor (p. 49)
- transportation (p. 3)
- U-shaped glacial trough (p. 49)
- visual pollution (p. 61)
- weathering (p. 3)
- wilderness (p. 54)
- World Heritage Site (p. 18)

SUMMARY

Having worked through this chapter, you should now know:

★ the characteristics of coastal landscapes, Carboniferous limestone landscapes and glaciated uplands in the British Isles, and the formation of their associated physical features

★ the factors involved in the evolution of these landscapes; structure, rock type, geology, major geomorphological processes, climate, drainage patterns, soils and human factors relating to changing land-use and settlement

★ the land resource base provides inter-related economic and social opportunities; agriculture (including the impact of UK and EU policies), forestry, recreation and tourism, mineral exploitation, industry and water storage

★ competing demands for land result in environmental problems and conflicts within these landscapes. There is a need for protected areas such as National Parks

★ UK and EU land-use policies can affect rural land use.

Section 2
RURAL LAND DEGRADATION

Links to the core

This chapter deals with the causes and consequences of degradation of rural land, and looks at the development of conservation and management techniques to prevent rural land degradation. Since the causes of degradation may be due to physical processes or human activity, it combines many of the ideas developed in both the Physical and Human Geography core. It builds on the concepts from the Atmosphere, Hydrosphere and Biosphere units and integrates aspects of population geography and rural geography into detailed case study material from North America and Africa.

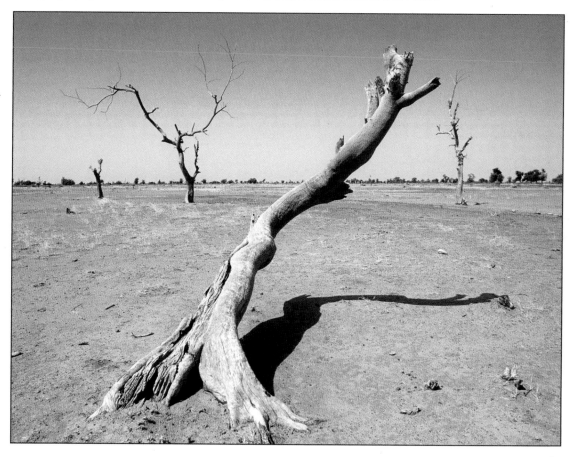

2.1 LAND DEGRADATION – WHAT IS IT?

Introduction

In a geographical context, rural land degradation is linked to soil deterioration. Soil quality is said to be deteriorated if it cannot be used in the way it was previously used, or it cannot support the vegetation cover it previously supported. The processes which cause this deterioration may be natural events, for example, floods or volcanic eruptions, or the result of human activity, or a combination of both.

Whatever the cause, we need to understand:

◆ what factors are responsible for the degradation of the soil/land

◆ the mechanics of the processes involved in soil erosion by water and wind

◆ the factors that determine how susceptible a soil is to erosion.

Land degradation has an enormous social and economic impact on the people living in affected areas. This chapter focuses on three case studies:

◆ the Tennessee Valley, USA

◆ the Great Plains of the USA

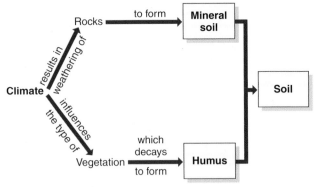

Figure 2.1 Soil formation

◆ the Sahel region, North Africa.

Three themes are followed in each case study:

◆ **The causes** – Why has land degradation occurred here? What physical and human factors are responsible? How do these factors work together to cause degradation?

◆ **The consequences** – What impact has degradation had on the landscape and on the people? What have been the social and economic consequences of land degradation?

◆ **The cures** – What has been done to prevent

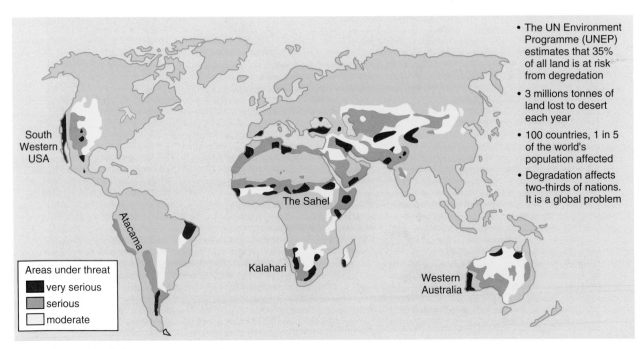

• The UN Environment Programme (UNEP) estimates that 35% of all land is at risk from degradation

• 3 millions tonnes of land lost to desert each year

• 100 countries, 1 in 5 of the world's population affected

• Degradation affects two-thirds of nations. It is a global problem

Figure 2.2 Global distribution of areas of degradation

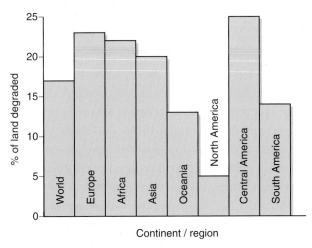

Figure 2.3 Human-induced soil degradation, 1945–90

soil will diminish, and, in certain circumstances, the cover of topsoil may be removed altogether, exposing the subsoil or parent rock. Land degradation is the result. It is most common in the tropics and subtropics, particularly in semi-desert environments. It can have far-reaching social and economic consequences for the people who live in affected areas, particularly in the developing world. Population growth is at its most rapid in the tropical and subtropical countries. More people require more food, and so the land is used more intensively. This is the main reason for the rapid soil degradation taking place in these areas (Figure 2.2).

further degradation? What soil conservation and land management strategies have been put in place?

Climate plays a major role in forming soil (Figure 2.1). The process can take hundreds of years before a mature soil is formed. Like any material on the Earth's surface, soils are exposed to the agents of **erosion**. If erosion happens faster than soil formation then the soil begins to degrade. The quality of the

Degradation is damaging agricultural land very quickly (Figure 2.3). Soils are under intense pressure as more food has to be produced on less land. Many farmers in the developing world are caught in a vicious downward spiral (Figure 2.4) which ultimately leads to degraded soils, low yields, low income, poverty and disease. Ironically, those areas which lose most soil are those that can least afford to do so. In the USA, 50% of fertiliser applied is used to

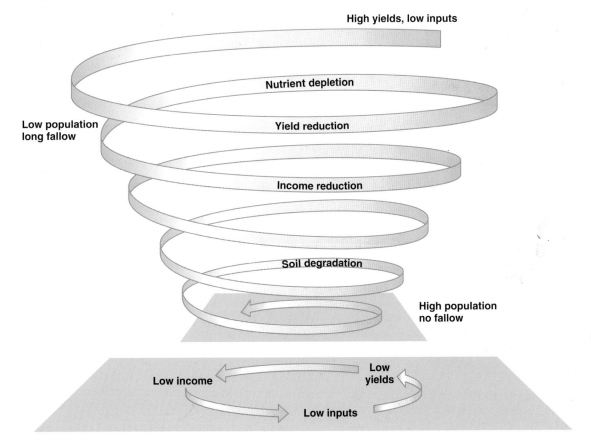

Figure 2.4 The downward spiral of poverty

compensate for loss in soil quality due to degradation. In Zimbabwe, Africa, loss of nutrients is three times greater than fertiliser input.

Such problems have led to the development of soil conservation and land management strategies ranging from capital-intensive, high-technology investments to low-technology, labour-intensive, small-scale, local self-help approaches.

The processes involved in soil erosion

A healthy soil is a living mixture of materials: inorganic particles such as sand or clay, decayed organic material (humus), air, water, micro-organisms, fungi and insects. All life in the soil is essential to maintaining fertility. Organisms mix nutrients through the soil, help ventilate it, add nutrients through animal waste and keep the structure crumby, so it is capable of holding moisture. Damage to soil life disturbs this ecosystem, and can lead to degradation and erosion.

There are three factors responsible for soil erosion:

◆ physical action

◆ chemical action

◆ biological action.

The impact of human activity on the land intensifies these actions, contributing to accelerated land degradation. Table 2.1 shows the human activities that have led to soil degradation across the world. Soil degradation in the developing world is a result of human activities such as deforestation and overgrazing (both brought on by population pressure). In the developed world, agricultural mismanagement is a much more important factor in land degradation. It is predominantly human activities that set in motion the various physical processes, for example, erosion by wind or running water, which in turn cause the soil to be degraded or stripped from the land.

Biological action is usually the result of removal of vegetation cover. Vegetation cover protects the soil in many ways, for example, roots help bind the soil, foliage intercepts rain thus reducing raindrop impact, and the presence of vegetation protects the soil from wind.

Removal of vegetation by humans or by natural processes exposes the soil to wind and rain and increases the likelihood of soil erosion.

Farming activities can lead to both **physical** and **chemical erosion**. Persistent trampling by animals or humans and regular use of heavy machinery compact the soil and damage its structure. Root growth is held back and water is unable to drain

Table 2.1 Causes of soil degradation (in million hectares of affected areas)

	Deforestation	Overgrazing	Agricultural mismanagement	Overexploitation	Bio-industrial activities
Africa	67	243	121	63	+
Asia	298	197	204	46	1
South America	100	68	64	12	–
North and Central America	18	38	91	11	+
Europe	84	50	64	1	21
Australasia	12	83	8	–	+
World	**579**	**679**	**552**	**133**	**23**

Source: Oldeman et al, 1990

freely through the soil, resulting in increased overland flow and more risk of erosion by running water. Trampling and machinery destroy plant cover, leaving bare patches exposed to wind and rain. Monoculture and over-cultivation weaken soils by removing too many nutrients. The soil becomes less able to support plant life and its healthy crumb structure breaks down leaving it at risk to attack by wind and water. Modern farming adds chemicals to the soil by way of pesticides, fertilisers and herbicides. These contaminate and damage soil life – bacteria, worms and insects – resulting in a vicious circle of soil depletion. To maintain crop yields, more chemicals have to be applied and the vicious circle continues. Acid rain degrades soil quality, particularly if the soil is already acidic. Irrigation is often used to increase crop production, but if not properly managed it can cause **salinisation** of the soil. Unwanted salts rise to the topsoil, rendering it useless for agriculture.

1 a What is land degradation?
   b Study Figure 2.2. Describe the pattern of land degradation across the continents of the world.
2 Complete the table below. Under each heading add examples of actions which can result in soil erosion.

Biological actions	Physical actions	Chemical actions

It is estimated that, in the last 50 years, human activities coupled with physical processes have degraded soils on 43% of the Earth's vegetated land. The chief factors involved are:

◆ indiscriminate felling of rainforests

◆ **overgrazing** by livestock in semi-desert environments

◆ inappropriate agricultural practices in all parts of the world.

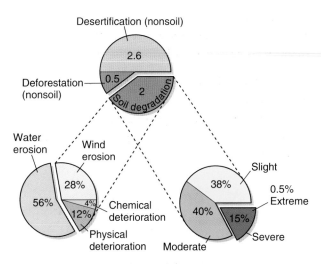

Figure 2.5 Global land degradation (billions of hectares)

Eroded soil must go somewhere. Damage is not restricted to the site of erosion; indeed, the environmental and economic damage suffered in places where the eroded soil is deposited may be just as severe. Soil particles blown or washed away can clog up rivers, lakes and reservoirs or smother neighbouring landscapes in a blanket of fine dust. Figure 2.5 shows the causes of soil degradation across the world and the level of severity of the problem.

Approximately 85% of soil degradation is a result of erosion by water and wind and there are four different levels of degradation:

◆ slight degradation – most soils could be fully restored easily

◆ moderate degradation – considerable financial investment is needed to restore soil quality

◆ severe degradation – currently soil is useless for agriculture

◆ extreme degradation – soil is incapable of restoration.

Soil erosion by water

There are four main types of soil erosion by water:

◆ **Rainsplash** is concerned with the impact of raindrops on the surface of a soil.

◆ **Sheet wash** is the removal of a thin, almost unseen, layer of surface soil.

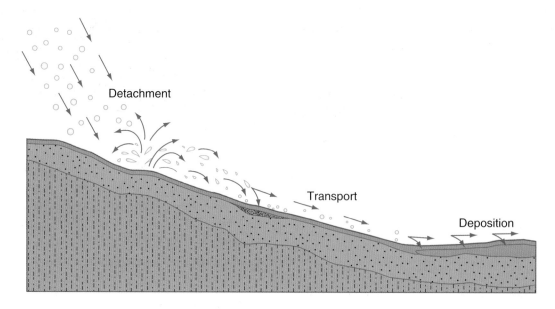

Figure 2.6 Three-step process of soil erosion

◆ **Rill erosion** is the creation of very small eroded channels across a soil surface.

◆ **Gully erosion** is the creation of large gullies by large quantities of water flowing over the soil surface.

Soil erosion by water can be summarised as a three-step process, as shown in Figure 2.6.

Step 1 Detachment – Soil particles are detached from the main body of soil mass. This can be done by raindrops hitting the soil or by overland flow of water across the soil surface.

Step 2 Transportation – Soil particles are carried downhill. They may float in water, roll or be dragged by water or splashed by raindrops.

Step 3 Deposition – Eventually the soil particles are deposited in a downhill location. It may be on the bed of a river or lake, or on the sea-bed.

Rainsplash

The bigger the raindrop, the faster it falls and the harder it hits the soil. Large drops, such as those in a thunderstorm, hit the soil at about 30 km/hr with an explosive impact creating a mini-crater. Soil particles are blasted outwards in a surrounding ring.

The force of rainsplash can be so great that:

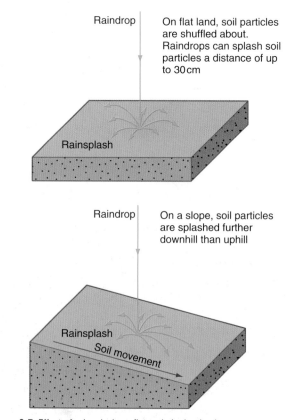

Figure 2.7 Effect of rainsplash on flat and sloping land

◆ particles of soil become loosened from the main body of soil mass

◆ soil granules are broken down into finer particles – the soil's 'crumb structure' is destroyed

◆ soil particles are exploded (splashed) at considerable distances from the point of impact, some as much as 2 m horizontally.

On a flat surface the impact of rainsplash is simply to move soil particles around in a random manner. However, if the soil is on a slope, then rainsplash will displace particles further downhill (Figure 2.7), resulting in the mass movement of soil downhill.

Sheet wash

The rainsplash effect can also clog up the soil, leading to a form of soil erosion called sheet wash (Figure 2.8).

The tiny soil particles which are displaced during rainsplash find their way into the soil pore spaces located between the soil crumbs and clog them up. The result is the formation of a surface crust (perhaps only 1 mm thick), but the rate at which rainwater can now infiltrate the soil is reduced. The result is more surface runoff which picks up the displaced soil particles and transports them downslope. The fragile crust survives until the next rainstorm when it is disintegrated by the first few raindrops, only to be re-formed again as rainsplash blasts a fresh set of tiny soil particles into the pore spaces.

Figure 2.9 Sheet erosion

If this happens on soils exposed to heavy rainfall where the slope is less than 5 degrees then sheet erosion will occur (Figure 2.9). Such slopes allow a very thin film of water to flow smoothly downslope (**sheet flow**). This process does not have the power to detach soil, but transports already detached particles.

Rill and gully erosion

These are the most serious forms of soil erosion by running water. Most land surfaces are irregular, with natural depressions, slopes and channels. A sudden burst of rainwater that is unable to soak into the soil will flow over the surface, finding its way into natural channels. Here it will gather in volume, power and speed as it makes its way downslope. The flowing water detaches particles of soil and moves them downslope.

The result may be numerous small, eroded channels criss-crossing the landscape, each no more than a few centimetres deep. These mini-channels are called rills. Rills are not permanent. Those formed in one rainstorm may be obliterated by the next storm, when an entirely fresh network of rills may be created, in positions unrelated to the previous network. They have no permanent connection to the drainage system. On farmland, annual ploughing will cover up and obliterate rills that were formed over the previous season, but the damage has already been done and the soil has been eroded (Figure 2.10).

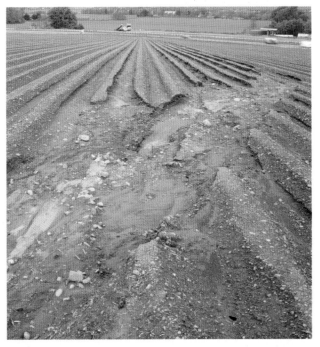

Figure 2.8 Erosion caused by sheet wash

Figure 2.10 Rill erosion

Figure 2.11a Gully erosion

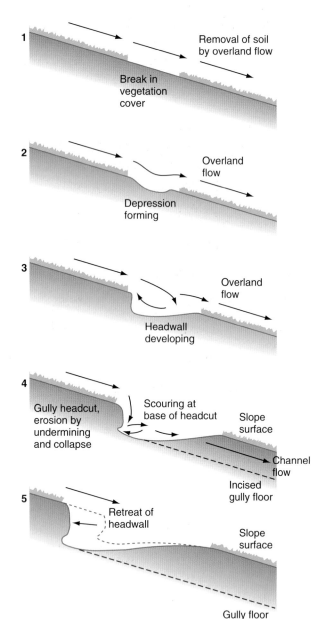

Figure 2.11b The development of gullies

Gullies are steep-sided water channels which carry water only during rainstorms and, unlike rills, are a permanent feature of the landscape. If the volume of runoff water becomes concentrated into one channel (on a slope of more than 5 degrees) the rushing water will cut deep into the soil. Gullies, which may be several metres deep and wide, are quickly carved out.

Gullies can develop as enlarged rills. They can also be formed where a break in the vegetation cover on a slope allows runoff to carve out a mini-depression. Over time this grows larger, allowing more rainfall to be trapped before making its way out of the depression to follow a channel (gully) downhill (Figure 2.11a). The gully can retreat uphill in much the same way as a waterfall retreats upstream. The headwall soil is progressively undercut, collapses into the channel and is washed downslope (Figure 2.11b).

Soil erosion by wind

Wind erosion is most common in arid or semi-arid areas. The following conditions encourage soil erosion by wind:

◆ soil that is loose, dry and finely grained

◆ a relatively flat land surface

◆ little or no vegetation cover

◆ large fields

◆ strong winds.

Strong winds blowing over relatively dry soils pick up the finer particles and may transport them hundreds of kilometres across continents and oceans. The topsoil, the most fertile portion of the soil, is literally blown away. It is a worldwide problem often exacerbated by human misuse of fragile semi-arid ecosystems, and often brought about by population pressure.

As with water erosion, erosion by wind is a three-step process:

◆ detachment

◆ transportation

◆ deposition.

The finest soil particles are transported in **suspension**. Wind picks them up, lifts them into the air (from a few metres to several kilometres high),

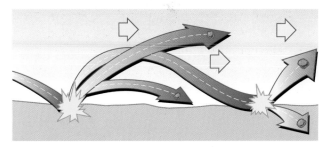

Figure 2.12 Process of saltation

and may blow them horizontal distances of several hundred kilometres. Dust storms are a spectacular visual indication of soil particles in suspension. These particles return to the ground only when the wind drops sufficiently or when rain washes them down.

The movement of larger particles by the wind by a series of short bounces or jumps close to the ground is called **saltation** (Figure 2.12). Saltation acts on particles that are too big to be carried in suspension in the air, but small enough to be picked up and raised off the ground by gusts of wind. They hit and move other particles as they travel.

The force of the wind may cause the largest particles to roll or slide along the soil surface in a process called **surface creep**. Smaller bouncing particles encourage surface creep when they hit the larger particles.

Although suspension is the most spectacular process of soil transportation it usually accounts for only about 15% of total movement; surface creep accounts for around 15% and saltation 70%.

Factors affecting soil erosion

The extent to which any soil is likely to be eroded by water or wind is a combination of a number of physical and human factors (Table 2.2). The rate at which soil is eroded is a function of both **erodibility** (the resistance of soil to detachment and transport) and **erosivity** (the potential of slope processes to cause erosion). A light, dry, sandy soil with limited vegetation cover on a steep slope in an area of tropical downpours would have high erodibility and erosivity, whereas a fine-grained clay on flat land with a dense vegetation cover in an area of gentle rainfall would have low erodibility and erosivity.

Table 2.2 Factors affecting soil erosion

Unlikely	Soil erosion	Likely
LOW	rainfall intensity	HIGH
LOW	runoff volume	HIGH
LOW	wind strength	HIGH
HIGH	soil infiltration rate	LOW
DENSE	vegetation cover	NONE
GENTLE	slope angle	STEEP
LOW	population density	HIGH
GOOD	soil and land management	POOR

Source: After Morgan, 1991

QUESTIONS

3 a Describe the three-step process that is common to all types of soil erosion.
  b With the aid of a diagram, describe the rainsplash effect.
  c Describe the conditions necessary for sheet wash to take place and explain the process.
  d Complete the table opposite to compare rills and gullies. Add bullet points to highlight similarities and differences in their appearance and formation.

Rills	Gullies

4 Describe and explain the differences between these different forms of transportation of soil particles by wind:
  ◆ suspension
  ◆ saltation
  ◆ surface creep.

Table 2.3 Percentage of total rainfall penetrating a vegetation canopy

Planting Density plants/m^2	Corn	Soyabean	Oats	Peas
25	57%	65.9%	78.4%	78.9%

Source: After LAL, 1988

Plant cover (vegetation)

The more dense the vegetation cover, the greater the protection the soil has from erosion. Just how much protection plants offer depends on the density of the ground cover, the density of the root systems and the height of the plants. Dense root systems bind the soil particles tightly together. A dense ground cover intercepts raindrops, absorbing their energy, thus protecting the soil underneath from the full force of impact. However, if the plant cover has a high canopy, as, for example, in a deciduous forest of beech and oak trees, small raindrops collect on the leaves, join up to form large drops and fall 20–30 m to the ground, thereby creating a significant rainsplash effect.

The most protective plant cover has a dense root system and a low canopy (under 1 m tall). Table 2.3 gives an indication of the percentage of rainfall which penetrates the canopies of various crops.

Figure 2.13 Effect of vegetation on soil loss

Figure 2.14 Effect of vegetation on wind velocity

Clearly, some crops offer better protection from rainsplash than others.

The relationship between plant cover and soil loss from water erosion is not a directly proportional relationship. Notice how, in Figure 2.13, even a small percentage of plant cover has a big impact on protecting the soil: a cover of only 10% gives 50% protection.

Vegetation also gives protection by protecting the soil surface from the effects of wind. On a bare, uneven soil (Figure 2.14), wind speed registers zero at ground level. However, the presence of plants significantly increases the zone of 'no wind' to about 70% of the height of the plant. Again, a small percentage of plant cover will give a relatively large degree of protection.

Rainfall and relief

The intensity of rainfall is an important factor influencing erosion. Intense rain has large drops which break up the soil crumb structure and there is more **surface runoff** to transport detached soil particles downslope. Gentle rain of low intensity causes little erosion, but torrential downpours may result in severe erosion – so much so that a handful of storms over a ten-year period may account for most of the soil erosion in that time. Generally, the most erosive rains occur in tropical and subtropical climates.

The steepness of a slope affects erosion much more than the length of the slope. Slopes most likely to encourage erosion are steep, short and convex. Erosion will be less on long, gentle, concave slopes, and negligible on flat land. The impact of slope on wind erosion is complex but soils most at risk are those on flat, open, treeless landscapes where the wind has a long fetch.

Soil type and structure

The rate at which water drains down through the soil – its **infiltration rate** – is another important factor affecting erosion. A coarse sandy soil with large particles and pore spaces has a high infiltration rate, whereas a clay-based soil with fine particles and small pore spaces has a low infiltration rate (Figure 2.15). A high infiltration rate means less water is available for surface runoff, and so there is less water erosion.

The structure of the soil is also important. Larger particles are less easily moved by water or wind than small particles. A coarse sandy soil is less likely to be blown or washed away than a fine-grained clay or silty soil (assuming the soils are dry).

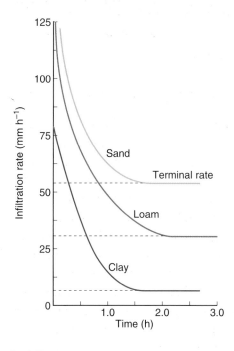

Figure 2.15 Soil infiltration rates

5 Construct a summary spider diagram which identifies all the factors affecting soil erosion.

6 Identify the ways in which vegetation cover protects soil.

7 Explain why:
  a short bursts of intensive rainfall are very effective in eroding soil

  b soils with a fast infiltration rate are less prone to erosion by water.

8 Study the table below giving details on three soils. Taking each soil in turn, assess the level of risk of soil erosion by water and explain your assessment.

Soil	Soil infiltration rate (cm per minute)	Vegetation cover	Rainfall
Soil 1	3.0	Close cover shrubs, small trees, grasses	500 mm per year, little seasonal variation, low intensity
Soil 2	0.5	No continuous ground over, grass clumps	300 mm per year, summer wet season, high intensity
Soil 3	1.0	Deciduous forest, sparse grass undergrowth	650 mm per year, little seasonal variation, low intensity

LAND DEGRADATION IN NORTH AMERICA

Tennessee Valley

Location

The area of study is the valley of the Tennessee River in east-central USA. The river is within the Mississippi River basin, draining much of central USA southwards into the Gulf of Mexico (Figure 2.16). The Tennessee River is extremely large. It rises on the western slopes of the Appalachian Mountains and flows westwards via a huge southerly loop, before joining the Ohio River, itself a tributary of the Mississippi. Although much of its 1000 km journey is within the state of Tennessee, a significant portion is in Alabama and its waters are also fed by streams originating in Georgia, North and South Carolina and Virginia.

Climate

Due to the sheer size of the river basin the climate can vary considerably. The climate graph for Knoxville (in the main valley floor in the upper part of the basin) is shown in Figure 2.17. On the higher ground in the Appalachian foothills, rainfall figures would be higher (up to 2000 mm per year) and

temperatures would be lower than for Knoxville. The climate graph for Glasgow is given for comparison, and the following points should be noted:

◆ The average temperatures are generally higher in Knoxville as it is much further south than Glasgow (Knoxville 36 °N, Glasgow 56 °N).

◆ The temperature range is greater in Knoxville. This is because Knoxville, unlike Glasgow, is far from any cooling influence of the sea in summer and warming influence of the sea in winter.

◆ The rainfall totals are broadly similar. This disguises a very important feature of rainfall in the Tennessee Valley. Throughout the year, but especially in summer, rainfall comes in heavy thunderstorms as warm, moist, unstable tropical maritime air pushes northwards from the Gulf of Mexico. Short, intense rainstorms on any unprotected soil on steep slopes will result in rapid soil erosion.

Soils and vegetation

The natural vegetation of the Tennessee region is forest: deciduous forest on the lower-lying land and

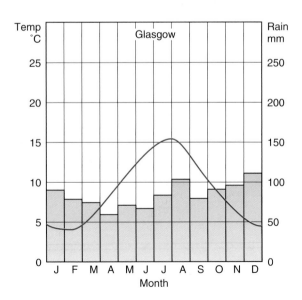

Figure 2.17 Climate graphs for Knoxville and Glasgow

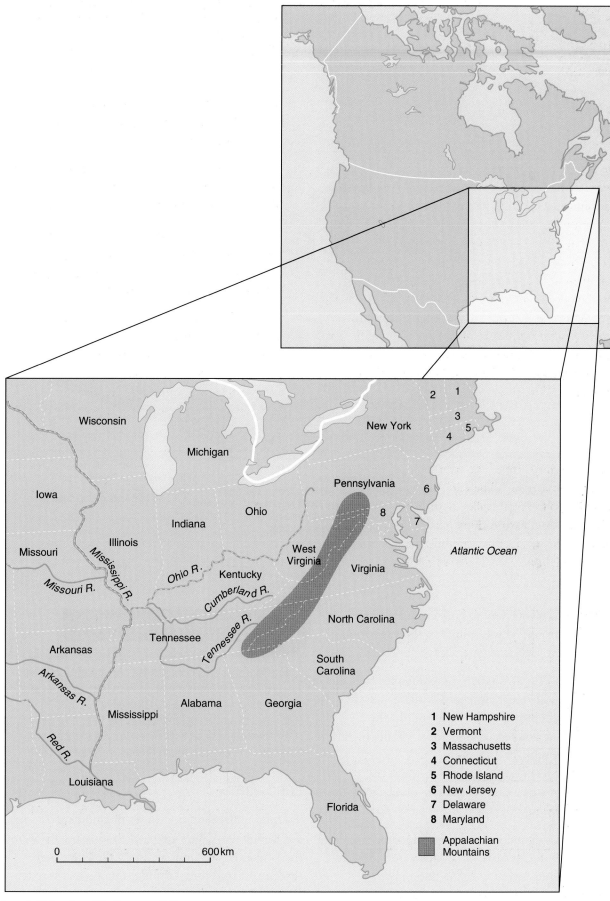

Figure 2.16 Location of the Tennessee Valley

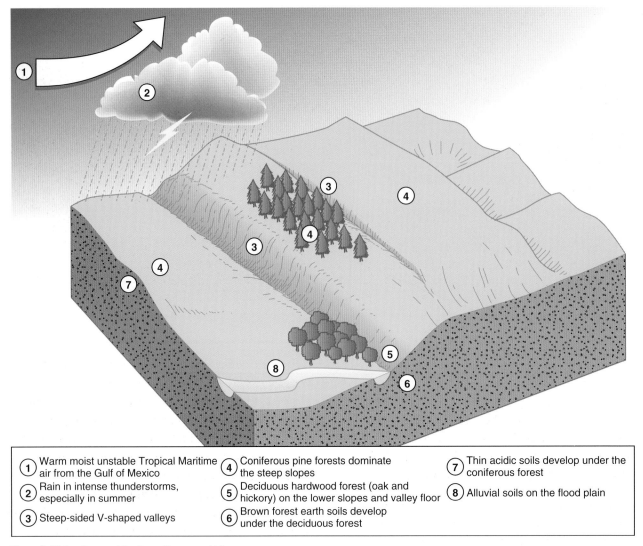

① Warm moist unstable Tropical Maritime air from the Gulf of Mexico
② Rain in intense thunderstorms, especially in summer
③ Steep-sided V-shaped valleys
④ Coniferous pine forests dominate the steep slopes
⑤ Deciduous hardwood forest (oak and hickory) on the lower slopes and valley floor
⑥ Brown forest earth soils develop under the deciduous forest
⑦ Thin acidic soils develop under the coniferous forest
⑧ Alluvial soils on the flood plain

Figure 2.18 Landscape of the Upper Tennessee Valley

valley floors, giving way to coniferous forest as altitude increases. This is similar to the vegetation pattern in the UK.

The pattern of soils is much more complex, but the main zonal soil is that associated with deciduous forest, namely brown forest earth soil. In the areas dominated by coniferous pine forest, on the high ridges and mountain tops and in the V-shaped valleys, the steepness of the land prevents the formation of a true podsol (the soil associated with coniferous forest).

In the main valley floor, successive flooding by the Tennessee River has left a thick carpet of alluvial deposits spread over the flood plain (Figure 2.18).

QUESTIONS

1 Compare and contrast the climate of Scotland (Glasgow climate graph) with that of the upper Tennessee River valley (Knoxville).
2 Describe the differences in the nature of rainfall between these two areas. Why is the potential for soil erosion much greater in Tennessee?

Land use

To fully understand the causes of land degradation in the Tennessee Valley we need to look briefly at how the land was used until about 1930. There were three main land uses: farming, mining and forestry (Figure 2.19).

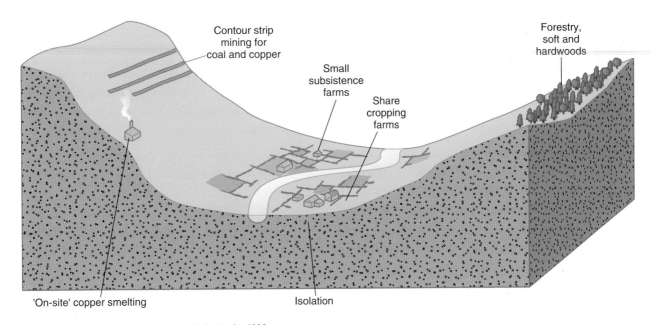

Figure 2.19 Occupations in the Tennessee Valley in the 1930s

Farming Over 50% of the population lived on farms and, of these, 25% lived on farms they did not own. Farming was mostly self-sufficient and confined to the lower slopes and valley floor (steeper slopes were forested). In some places soils were too poor to farm, while in others, for example, in northern Alabama, the fertile soils supported cotton farming.

Mining Mining for coal took place in an area called the Cumberland Plateau, to the west of Knoxville. Coal seams were thin but close to the surface so mining was carried out in opencast strips which followed the contours of the landscape. This left ugly scars on the land. Mining for copper was carried out east of Chattanooga (at Copperhill on the Tennessee–Georgia border). Forests were cut to fuel copper smelting which was done 'on site'.

Forestry Forestry was a significant employer, especially in the Blue Ridge area of North Carolina and in the Highland Rim area south-west of Nashville. By 1930, lumber companies had logged most of the natural forest of softwoods and hardwoods in both of these areas, leaving the soils, often on steep slopes, exposed to heavy rainstorms.

Social and eonomic conditions

The Tennessee Valley was a pocket of poverty and social deprivation in the 1930s. Incomes were well below the national average. Poverty was reflected in health: diseases such as TB, typhoid and malaria shortened life expectancy, malaria affected 30% of the population and malnutrition was widespread. Birth rates were high (30% higher than the national average) as were infant mortality rates. Access to health care services was very difficult, if not impossible, for people on such low incomes. Education was poor – illiteracy rates were double the national average – and much of the valley basin remained an isolated 'backwater', essentially cut off from the rest of the country by a skeletal communication network. The river, which might have provided a communication spine, was made useless for navigation by regular flooding along most of its length. The river regularly changed channel and in places became very wide and very shallow, less than 0.5 m deep.

These factors provided major disincentives to industrial development. As the USA industrialised in the early years of the twentieth century, Tennessee was ignored. Social and economic conditions became so harsh that some mountain villages were completely abandoned. Other villages and towns lost many of the younger people as they moved away to find work and a new life elsewhere in the USA.

The causes of land degradation in the Tennessee River basin

Land degradation in Tennessee was caused by a combination of physical and human factors.

Physical causes

The landscape of the upper Tennessee River basin is vulnerable to land degradation through soil erosion. The reasons for its vulnerability are the numerous steep-sided V-shaped valleys of the tributary streams and the nature of the rainfall, as short heavy downpours. The role of vegetation is crucial here. The foliage and root systems of the dense coniferous forests in the upper basin protect the soil. Root systems help bind the thin soil together and foliage acts as a shield, protecting the soil from rainsplash impact during the heavy downpours. Unprotected soil is very vulnerable and quickly carried downslope by sheet wash, rill and gully erosion.

Human causes

The human causes of land degradation are bad mining practices, bad forestry practices and, most significantly, bad farming practices.

Strip mining for coal left ugly open scars on the land. Trees were cleared leaving a bare land surface exposed to attack by rainwater. Mounds of mining waste suffered sheet and gully erosion, with silt being

Figure 2.20 Eroded land

carried by fast-flowing mountain streams into the main river itself. Copper ore was smelted on site with local wood being used as fuel. The newly exposed deforested soils were quickly washed from the hillsides. Poisonous fumes from the smelting process killed nearby vegetation, while poisonous sulphuric acids from the smelters were carried into the river.

Logging companies stripped the hillsides quickly and indiscriminately, so that by 1930 almost 50% of the original tree cover had been removed. Little attention was paid to proper forest management practices and replanting, which would have protected the soil and ensured a supply of timber in the years to come. The thin soils on the steep slopes no longer had the protection that vegetation cover offered, and heavy rains washed the soil into streams.

The worldwide economic recession of the 1930s saw mass unemployment in the factories in the industrial heartland of north-eastern USA. As a result, former Tennessee emigrants drifted back to their homelands in the valley. New land on steep slopes was cleared to create new farms. Farming was barely above subsistence level, and farmers would abandon fields which had low yields, clear forest and try again, leaving the abandoned land to soil erosion. In total, about 15 million hectares were cleared for farming. Of these, 3 million needed erosion control measures and a further 0.4 million were eroded to the point of abandonment (Figure 2.20).

In addition, various bad practices employed by farmers contributed significantly to soil erosion:

- **Ploughing** Ploughing up and down the slope created furrows which became ready-made channels transporting the rainwater downslope. In heavy storms the soils could not absorb the rainwater quickly enough so excess water ran downhill using the furrows. Rills and then gullies were quickly carved out as the soil was washed away.

- **Monoculture** The repeated planting of the same crop year after year resulted in a depletion of nutrients in the soil leading to a breakdown in soil structure, making it easier for wind and rain to carry soil particles away.

- **Fertilisers** At this time chemical fertilisers were

Figure 2.21 Causes of land degradation in the Tennessee Valley

not widely available and buying them was far beyond the means of most farmers. The application of available manure could not properly restore the soil. Soils became exhausted quickly and were very susceptible to erosion by wind and rain.

♦ **Soil exposure** The widespread practice of leaving the soil unprotected for part of the year led to sheet erosion becoming a common problem. This problem was particularly severe in the cotton growing lands of northern Alabama.

These causes of land degradation are summarised in Figure 2.21.

QUESTIONS

3 a What were the three main land uses in the Tennessee Valley in the 1930s?
   b Make a bullet point list of the social and economic conditions experienced by people in the Tennessee Valley in the 1930s.
   c Explain why industry was reluctant to move to the Tennessee Valley at this time.
4 What were the physical causes of land degradation?
5 'Bad practice' lay at the heart of the human causes of land degradation. Construct two spider diagrams, one to show bad farming practices, the other showing bad mining practices.

Consequences of land degradation

The result was massive soil erosion right across the Tennessee basin. Vast quantities of soil were stripped from the land and transported into fast-flowing mountain streams. These streams channelled the soil and silt into the much slower-flowing Tennessee River. Here it sank to the river bed, clogging up the channel and forming numerous sand bars and mud banks (Figure 2.22). In places the river became extremely shallow and very wide as water spilled out of the clogged channel onto the flood plain. Flooding was a regular annual occurrence, sometimes twice a year, across the entire valley downstream of Knoxville.

Regular flooding had a devastating impact on the people and the economy in the Tennessee Valley. The river was unmanageable and its shallow depth and shifting sand bars made it useless for navigation. People living on the flood plain suffered most. This area contains the best-quality farmland in the region, but each time it flooded farmers suffered ruined crops and lost harvests. This forced some farmers to move further up the valleys, clear new land and try to make a new start.

There was little incentive for investment in industrial development in the region for several reasons.

♦ Communications were poor. Roads and bridges were often closed due to floods. The river, a potential source of cheap transport, was

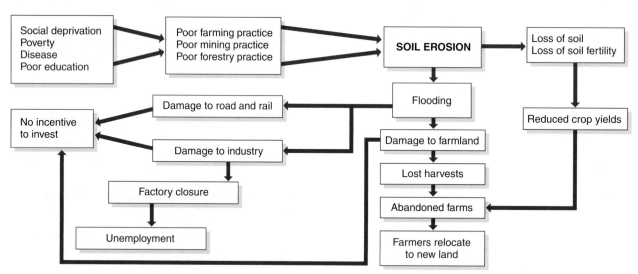

1 Soil and silt from bad forestry practice
2 Soil and silt from bad farming practice
3 Soil and silt from bad mining practice

Figure 2.22 River channel and sand bars

Figure 2.23 Causes and consequences of land degradation

unnavigable. Railway companies were reluctant to invest in new track.

◆ The only power source available in the quantities required by industry was coal. It was bulky and expensive to transport, and waste had to be disposed of after burning.

◆ The people in the region suffered from poor educational standards and poor health. A large

skilled workforce would not be available without investment in education and training.

The root cause of all these problems was soil erosion. If this could be controlled, then the river could be managed, investment attracted and the social and economic recovery of the area could begin. Figure 2.23 summarises the causes and consequences of land degradation.

QUESTIONS

6 a Explain how soil erosion in the upper Tennessee valley caused flooding downstream.
 b Summarise the consequences of flooding for:

◆ farming
◆ industry
◆ communications
◆ people.

Tackling erosion in the Tennessee Valley

An integrated plan was needed to solve the complex network of related problems in the Tennessee Valley: soil erosion, flooding, social deprivation, isolation, unemployment and a degraded landscape.

In 1933 the National Recovery Administration was set up with a budget of US $3 billion to spend on public works to get people back into employment. Part of this budget was spent on the creation of the Tennessee Valley Authority (TVA) in 1933. The TVA was originally given limited tasks:

TVA Act 18 May 1933
'To improve the navigability and to provide for the flood control of the Tennessee River; to provide for reforestation and the proper use of marginal lands in the Tennessee Valley; to provide for the agricultural and industrial development of the said valley'.

However, the TVA quickly and successfully expanded its remit to incorporate many other tasks:

◆ implement soil conservation strategies

◆ stem rural depopulation

◆ stimulate industrial regeneration

◆ improve the social and economic well-being of the people.

This was the first US example of large-scale government planning covering a complete river basin stretching across seven different states. The following text illustrates some of the schemes implemented by the TVA.

Reforestation

This was urgently required in the upper valley of the Tennessee River basin to stem the flow of eroded soil. Biological research suggested that the fast-growing loblolly pine was the best tree to hold and trap soil. TVA tree nurseries were set up to provide a supply of trees, and millions of hectares were replanted on government lands over a 50 year period. Trees were provided free to private landowners so that they could plant shelter belts and reforest marginal hill land. Education programmes were delivered to farmers providing information on good forest management practices. In addition to

aiding soil conservation, the forests also improved water quality, regulated its flow, provided opportunities for recreation, created new wildlife habitats and formed a major scenic resource.

Reclaiming the mining landscape

Reclamation work on numerous abandoned mines was carried out by the TVA. This involved covering mined strips with soil and planting with pines – approximately 2500 trees per hectare at an average cost of US $865 per hectare (1965 prices). Since 1965 mining companies have had stringent reclamation requirements written into their contracts. The treatment of toxic metals and acids is paid for by imposing a tax on mining companies proportional to the volume of minerals excavated. New mining techniques called 'cut and fill' which restore the landscape after mining are now widely practised.

Improving farming practices

Approximately 15 000 demonstration farms were set up across the TVA area. These served farmers by providing a local, easily accessible, open air classroom where the results of good farming practice could be seen and be evaluated. In addition the TVA ran an 'Agricultural Credit' scheme. This allowed farmers to borrow money cheaply to improve their farms and invest in new technologies. Further assistance was offered through the Rapid Adjustment Farming Programme. This provided any farmer with free individual advice on which crops, livestock and land management techniques best suited the conditions on his land. 'Electro Development Farms' illustrated how the use of electrical equipment could save time and money while increasing productivity and living standards.

A number of soil conservation techniques were introduced on the demonstration farms, including the following.

◆ **Contour ploughing** on the gentle to moderate slopes – this means that rainwater, which would otherwise run downslope causing the soil to be stripped away, is held in the furrows allowing time for it to soak into the soil.

◆ **Strip cropping** on very gentle slopes and flat land –

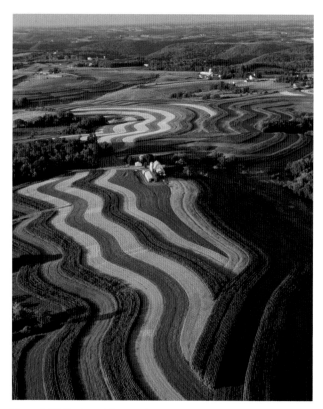

Figure 2.24 Contour strip cropping

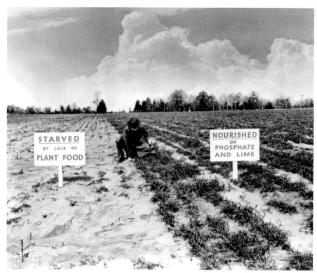

Figure 2.25 Fertiliser demonstration plot

tall crops are grown in narrow strips alternating with ground-hugging crops. If water flows downslope the ground-hugging crop will hold it back, giving it time to soak into the soil. The taller crop will also 'lift' any wind off the ground thereby reducing the likelihood of wind erosion (Figure 2.24).

- **Intercropping** – planting two crops together in the same field, but harvesting them at different times, ensuring the soil is covered at all times.

- **Double cropping** – two crops are grown in the same field in different seasons thus ensuring continuous ground cover. Wheat and soyabeans work well together.

- **Crop rotation** helps maintain soil fertility and structure by the alternate withdrawal and replacement of different minerals by different crops.

- **Chemical fertilisers** – government-run fertiliser stations conducted research and testing of fertilisers in order to produce 'tailor made' fertilisers to suit the needs of specific farms (Figure 2.25).

- **Crop changes** – farmers were encouraged to change to crops that were better suited to the soils on their farm. Crops which quickly exhaust the soil (like corn) went into decline while others (like soya) increased.

- **Runoff interception** – particularly vulnerable land was protected by constructing concrete-lined channels and ditches to intercept and control rapid **runoff**.

Flood control, HEP and industrial development

The risk of flooding was reduced by soil conservation strategies in forestry, mining and farming. Flood control, river navigation and the generation of cheap hydro-electric power (HEP) can only be achieved by building dams (Figure 2.26). A total of 30 dams were constructed on the Tennessee

Figure 2.26 Pickwick Landing Dam on the Tennessee River

River and its tributaries. The majority of these were small dams in the upper valleys and their purpose was mainly to regulate river flow but also to generate cheap HEP. Larger downstream dams generated HEP and also improved communications as roads were built over them and lock gates were constructed to allow river navigation.

Everything was now in place to attract large-scale industrial development:

◆ the risk of flooding to factories and communication links was gone

◆ navigation was now possible from the Gulf of Mexico via the Mississippi, Ohio and Tennessee rivers all the way to Knoxville – this offered cheap transport especially suited to the movement of bulk raw materials or manufactured goods

◆ cheap HEP was available in vast quantities

◆ there was a ready supply of labour – regular, well-paid work was attractive to hard-pressed farmers.

Large-scale manufacturing industry took advantage of these conditions and moved in. Aluminium smelting utilised local supplies of bauxite, cheap HEP and

cheap transport, as did iron and steel manufacturing, car assembly and the arms industry. In their wake came jobs through numerous supply and support factories and service industries.

Rural communities benefited from the availability of cheap HEP. The Rural Electrification Administration brought electricity to villages and farms for the first time (Figure 2.27). Farmers were able to invest in new technologies, helping them to improve productivity and increase efficiency. Electrified farms increased their income by up to 200% in five years, the biggest single benefit being the introduction of refrigeration. The amount of farmland was reduced, however, as land was flooded by the creation of new reservoirs. Farmers on the more marginal hill lands abandoned the land to take advantage of regular well-paid work in the new factories in the lower valley.

Effectiveness of the TVA

The TVA successfully implemented an integrated plan to solve the complex network of inter-related problems in the region. It put in place a range of strategies which conserved soil, controlled flooding, stimulated agricultural and industrial growth, stemmed rural depopulation and greatly improved the social and economic well-being of the people. As an exercise in large-scale, government-led social, economic and environmental regeneration it is an outstanding success. Its work is not over. Recently, soil erosion problems resurfaced when some farmers were tempted to overwork the soil in pursuit of large profits from growing soya beans. Programmes are still in place to continue research and farmer education in soil conservation techniques, to monitor soil erosion and to reclaim damaged land. A spin-off benefit from all of this work has been the development of recreational facilities. Forests, reservoirs and rivers provide the environment for numerous forms of outdoor activities, bringing more employment opportunities and income to the valley. The area now caters for five million visitors per year. Figure 2.28 summarises the main changes brought to the valley by the TVA.

Figure 2.27 Installation of electricity lines

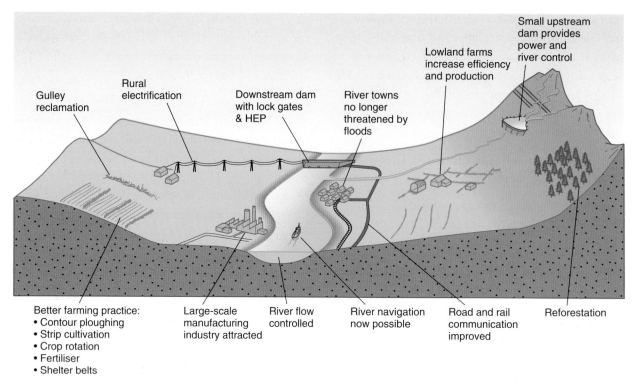

Figure 2.28 Changes brought to the Tennessee Valley by the TVA

QUESTIONS

7 a What were the main elements of the TVA's integrated plan?

b Describe the steps that were taken to reforest the Tennessee Valley.

c Mining companies must now clean up their own mess. Describe how this is done.

d Draw a spider diagram to summarise the programmes and techniques used to improve farming practice.

e Explain why flood control was crucial in attracting industrial development to the Tennessee Valley.

8 How effective has the TVA been in achieving its aims? Explain your answer.

Decision making question

On 7–8 January 1998 heavy rain fell on the drainage basin of the Doe River in Tennessee (the Doe is a tributary of the Tennessee River). Figure 2.29 shows the hourly histogram of rainfall at Burbank in the

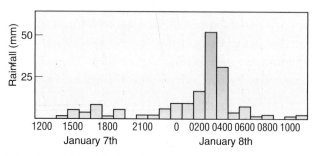

Figure 2.29 Rainfall in Burbank, Tennessee, January 1998

Figure 2.30 Level of the Doe River, 7–8 January 1998

river basin and Figure 2.30 shows the level of the Doe River over the same time period at Elizabethton, 16 km downstream from Burbank.

◆ Comment on the relationship between the pattern of rainfall and the level of the river over the time period shown.

◆ Comment on and suggest reasons for the length of the time lag.

◆ Construct an interlinked flow diagram to show how elements of the TVA scheme are connected. Use the example provided here as a starter.

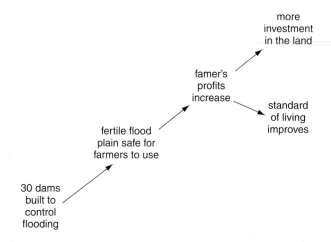

CASE STUDY
The Great Plains

The Great Plains cover an area of approximately 36 million km² extending for 3870 km north to south and 1612 km west to east (Figure 2.31). With a population of approximately 10 million people, population density is extremely low. This region supplies around 25% of the world's wheat, oats, sorghum and corn. Rural land degradation in the Great Plains is a combination of physical, climatic and human factors. It is estimated that 1.7 million tonnes of topsoil are either blown or washed away in the USA each year.

Climate

Owing to the wide range of latitude, climate varies across the Great Plains but in general terms it can be described as continental. As Figure 2.32 shows, annual precipitation totals are moderate (400–800 mm) while the annual range in temperature is high (−10 °C to 30 °C). Winters are long, dry and very cold. Owing to the low temperatures snow is common and can lie for up to five months. The summer months are hot, with temperatures often exceeding 30 °C. In July and August precipitation is heavy but variable. Evaporation rates are high, often

resulting in a moisture deficit, but this is offset by snowmelt in early spring.

Natural environment

The Great Plains are predominantly an area of undulating grasslands. The density and range of species of the vegetation vary according to altitude, latitude and climate. The natural ecosystem of the Great Plains is diverse. The fertile soil supports a rich grassland cover as well as a wide variety of insect life and biota. The dominant herbivore was bison, which grazed the grasslands.

Chernozem is one of the main soil types found in the Great Plains (Figure 2.33). It is one of the most

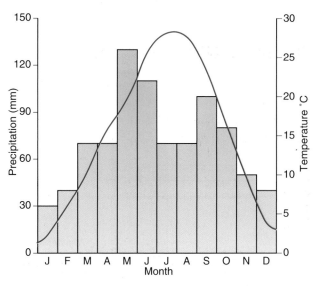

Figure 2.32 Climate graph for Oklahoma City

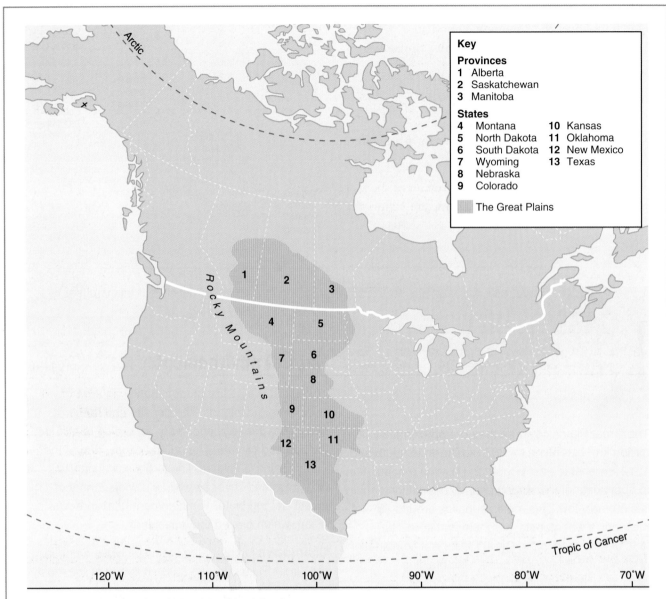

Figure 2.31 Location of the Great Plains of the USA and Canada

Key
Provinces
1 Alberta
2 Saskatchewan
3 Manitoba
States
4 Montana   10 Kansas
5 North Dakota 11 Oklahoma
6 South Dakota 12 New Mexico
7 Wyoming   13 Texas
8 Nebraska
9 Colorado

The Great Plains

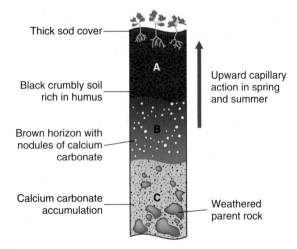

Thick sod cover

Black crumbly soil rich in humus

Brown horizon with nodules of calcium carbonate

Calcium carbonate accumulation

A

B

C

Upward capillary action in spring and summer

Weathered parent rock

Figure 2.33 A chernozem soil profile

fertile soils in the world. It is black, has a deep humus layer and crumby texture which makes it both fertile and easily cultivable. The decomposition of the grass provides the rich humus layer.

Rainfall increases from west to east, and enables the region to be divided into three broad vegetation bands:

◆ the short-grass prairie in the rainshadow of the Rockies
◆ the mixed-grass prairie in the central Great Plains
◆ the tall-grass prairie in the wetter areas to the east.

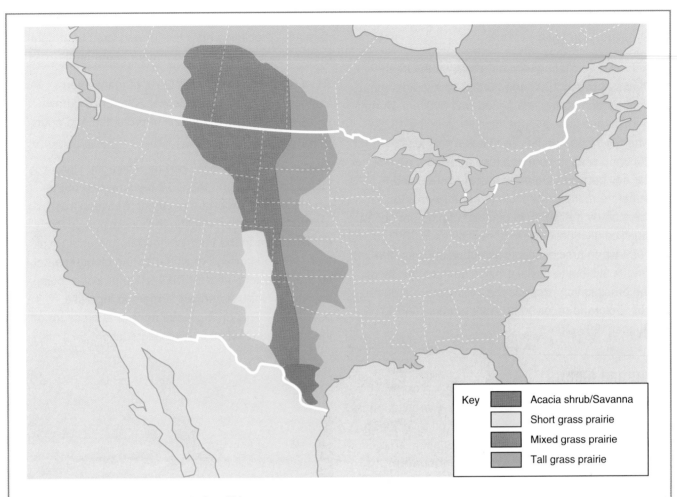

Figure 2.34 The natural vegetation of the Great Plains

These three zones broadly correspond to the western Rangelands (largely livestock farming), the Wheat Belt and corn/soyabean areas respectively (Figure 2.34).

Human activity

The traditional hunter/gatherer lifestyle of the Native Americans was in harmony with the Great Plains ecosystem. It caused no long-term damage to animal numbers, soil or vegetation. However, as North America was colonised, migrants moved west and the agricultural potential of the Great Plains was discovered. Settlers established cattle ranches, forcing the Native Americans off their homelands and slaughtering the majority of the bison herds.

By the late nineteenth century the settlers had realised the great potential in the Great Plains for the cultivation of wheat. The low-lying relief, fertile soils and grassland vegetation led farmers to believe that the Prairies were ideally suited to arable farming. Initially the switch to farming was very successful but the farmers failed to take into account the variability of rainfall in the area. This, combined with several other factors as outlined below, led to severe land degradation, which North America is still dealing with.

The causes of rural land degradation in the Great Plains

Rural land degradation in the Great Plains cannot be attributed to any one factor but is a combination of physical, human and economic conditions.

Physical factors

Annual precipitation varied greatly in the Great Plains between 1900 and 2000. The average annual precipitation during this period was 390 mm. During periods of below average rainfall, as in the 1930s when average annual precipitation was less than 200 mm, soils dry out and lose adhesion. As a result the soil particles are more loosely bound and therefore more susceptible to **aeolian erosion** (erosion by the wind). In contrast, during periods of significantly above average rainfall, as in 1995 to 1999 when precipitation was above 600 mm, the land is subject to fluvial erosion as the intensity of the precipitation exceeds the infiltration rate of the soil. Colorado received 100 mm of rainfall on 4 August 1999 alone!

Human factors

The success of extensive commercial farming in the Prairies depends on growing wheat on a very large scale. However, the cultivation of the same crop on the same fields year after year (**monoculture**) depletes the soil of vital nutrients. As a result the soil structure is weakened, crop yields are reduced and the land degraded. In addition, the root network of the natural grassland is much denser than that of the wheat and so under cultivation the soil becomes more loosely bound.

In preparation for planting, the fields are cleared and ploughed. This breaks up the crumby structure of the soil and leaves the fine, fertile topsoil particles exposed to wind erosion. Periods of low rainfall and high temperatures enhance the loss of soil as it dries out. Ploughing produces furrows in the fields. When it rains, the water is channelled into the furrows and runoff, carrying soil with it, is increased. This is a particular problem where the furrows are cut at right-angles across the natural contours of the land.

As herd sizes increased in parts of the western Prairies, natural grasslands were overgrazed down to their roots, leaving the soil unprotected and open to wind erosion.

The growth of extensive, commercial agriculture was highly profitable, and the USA became one of the world's largest exporters of wheat. Between 1914 and 1917, wheat production in Russia was severely depleted due to the First World War, and as a result there was a shortage of wheat on the global market, so the price rose sharply. This led to pressure being applied to North American farmers to increase productivity. This drive placed additional pressure on the land and resulted in land degradation.

Figure 2.35 The Dust Bowl

QUESTIONS

9 What physical and climatic characteristics of the Great Plains make it an area that is susceptible to rural land degradation?

10 What lessons, if any, should the new settlers in the Great Plains have learned from the traditional way of life practised by the native Americans?

11 Were economic, climatic or agricultural factors the biggest contributor to land degradation?

Consequences of land degradation

Physical impacts

The most obvious and devastating impact of land degradation in North America has been the loss of topsoil. During the 1930s millions of tonnes of soil were blown across huge distances, smothering vegetation, choking wildlife, burying buildings and blocking out sunlight. The region became known as the **dust bowl** (Figure 2.35).

'Home was the safest place to be, but even that offered only limited sanctuary. The dust was so thick that it filtered into houses as if walls were made of cheesecloth, not stone. The rags stuffed under doors and windows could no longer hold it out. Sand swirled into closets and cupboards, leaving its grimy fingerprints on dishes and clothes, sparing neither porcelain bowls lovingly passed through generations nor Easter dresses purchased yesterday.'

'Art Leonard sat in the middle of his family's tire store, listening to the thrashing winds and gazing at the miniature dust storms billowing about the room. Those daring to peek through open doors were stung by sand on their faces and bare legs. Many would later talk about how as they peered out the windows, they saw Kansas, Oklahoma and little bit of Texas roll by.'
(The Discovery Channel; http://discovery.com/area/ history/dustbowl)

The loss of soil fertility caused by wind erosion led to an increase in the use of fertilisers and pesticides, as farmers fought to sustain crop yields. In the long term this has resulted in the contamination of soils, rivers and aquifers. This in turn has affected indigenous flora and fauna, raising concerns about the environmental impact of extensive commercial agriculture.

Social and economic impacts

The large-scale erosion and land degradation effectively wiped out the livelihoods of thousands of farmers and related services, for example, shopkeepers and grain merchants across the Great Plains. They were left with the stark choice of abandoning the area in search of alternative employment or staying and attempting to rebuild their businesses. Many of those who left migrated to cities such as Detroit, in search of jobs in the car industry, or to Los Angeles. Farms were sold cheaply or even abandoned. This pattern of events led to rural depopulation across the region, and an increase in average farm sizes as farms were amalgamated.

The social costs of land degradation were high. Many people suffered from depression, suicides were common and unemployment rose sharply. The US Government was faced with the burden of providing financial support to those affected. Economically the USA was hit very hard, as it lost the valuable revenue normally brought in by the export of grain to regions such as Europe.

Solutions to the land degradation

Afforestation

One of the most effective ways to reduce the risk of land degradation is the planting of trees, and in particular **shelter belts**. Given the low-lying relief of the region and the strong winds that blow across it,

Figure 2.36 Different formations of shelter belt

the creation of shelter belts along field boundaries can reduce soil loss dramatically. Research has shown that trees are able to slow down wind speeds for a distance equivalent to ten times the height of the trees. However, their effectiveness is dependent on the formation of the shelter belt, as shown in Figure 2.36. Although extremely effective and relatively cost-efficient, the main disadvantages of shelter belts as a technique are the length of time required for the trees to mature, the maintenance they require and the area of cropland taken up by the trees themselves.

Contour ploughing

By ploughing along the lines of the contours instead of across them, farmers can reduce the amount of soil lost to overland flow by up to 50%. Contour ploughing reduces the speed of water flow across the land surface and therefore the erosivity and carrying capacity of the runoff is reduced. In addition to this, infiltration rates are greater, and growing conditions are enhanced by the increase in soil moisture. The only significant disadvantage of this technique is the additional workload involved in the identification, marking out and ploughing of the contours.

Laser levelling

Economically developed countries like the USA and Canada are in the privileged position of being able to employ 'high-tech' solutions to the issue of land degradation. Laser levelling is one such technique

Figure 2.38 Minimum tillage

(Figure 2.37). The aim is to achieve as level a surface as possible in the fields. Lasers are used to control the height of rollers, which are towed across the surface of the soil. Where the laser detects an increase in the height of the soil surface the rollers are lowered and where a lowering of the surface is identified, the rollers are raised. The extremely flat surface produced means the effects of rainsplash and overland flow are minimised. However, the equipment involved is expensive.

Minimum tillage

The principle of minimum tillage is based on limiting the number of trips made by machinery across the surface of fields in preparing, planting and harvesting crops. An example of minimum tillage is the **sweep plough**. Traditional ploughs work by turning the soil over in preparation for the planting of the following season's crop. This process breaks up the crumby structure of the chernozem soil leaving it exposed to processes such as rainsplash, overland flow and aeolian erosion. The sweep plough has been designed to cut below the surface of the soil, severing

Figure 2.37 Laser levelling

the roots of the previous crop and any weeds, without disturbing the soil surface. The fact that the stubble is left standing means it is able to afford some protection to the soil, as the new crop grows. Over time the stubble will decay, returning nutrients to the topsoil. The practice of leaving crop debris lying on the surface of the land is known as **mulching**. Arable farmers now have access to machinery designed to sow seeds, spread fertiliser and add pesticides simultaneously (Figure 2.38). This cuts down on the number of trips the farmer needs to make across the fields, minimising soil disturbance and compaction.

Fallow period

In the drive to maximise yields, many farmers in the Great Plains cut down or eliminated altogether a fallow period from their cycle of cultivation. This resulted in soils becoming exhausted, yields falling and land becoming degraded. Organisations like the **Soil Conservation Service (SCS)**, set up by the US Government in the aftermath of the Dust Bowl, now encourage farmers to ensure that land is left fallow on a rotational basis.

Intercropping and strip cultivation

This normally involves the cultivation of two or more crops in the same field (Figure 2.39). One such combination practised in the Prairies is corn or soyabeans along with oats or legumes. The idea is to use crops that grow at different rates and/or are harvested at different times of year so that the field is never devoid of plant cover. Once again, this reduces

Figure 2.39 Intercropping

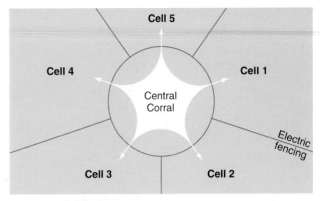
Figure 2.40 Cell grazing

the effects of wind and water erosion. In tests it has been found that in parts of the Great Plains intercropping has resulted in yields of between 10 and 40% greater than from cultivation of one crop.

Cell grazing

This system, used by ranchers in areas such as Colorado, is designed to combat the effects of overgrazing. The rancher divides his land up into cells which are arranged around a central corral (Figure 2.40). The cattle are restricted to one 'cell' at a time and allowed to graze that area for three days. They are then moved to another cell and the grass is given time to recover and rejuvenate.

This system is relatively simple but very effective, providing the rancher has sufficient land to create the number of cells necessary to make it work and the resources available to fence off the cells.

Bison ranching

Ranchers in some areas of the USA, such as Montana, are now rearing herds of bison instead of cattle in a bid to prevent overgrazing. Bison are one of the indigenous species of the Great Plains and in the wild they do not overgraze the natural grasslands. Ranchers have found that bison will graze and move on from one area to the next before the vegetation cover is irreplaceably damaged. The higher-yielding strains of beef cattle brought into the area tended to strip the land of all cover (including the roots) thus leaving it susceptible to wind erosion.

Figure 2.41 Boom irrigation in Colorado

Farmers are able to receive financial assistance for switching to bison, and consumers are being won over by schemes that promote the potential health benefits of eating bison meat which has 30% more protein and 25% less cholesterol than beef.

Boom or centre-pivot irrigation

If arable farmers can eliminate the potential risks of drought by irrigating crops then they are able to sustain crop cover and therefore minimise degradation from wind erosion. This can only be achieved in locations where a suitably large and sustainable source of water is available, such as a surface reservoir or underground aquifer. In areas such as northern Texas and New Mexico this has been made possible by extracting water from the Ogallala Aquifer. As Figure 2.41 illustrates, boom

irrigation has resulted in one of the most spectacular farming landscapes in the world. However, this technology comes at a cost. Not only is the equipment required to pump and distribute the water very expensive, but in many areas the rate at which water is being removed from the aquifers greatly exceeds the rate at which the water is replaced by precipitation. In time this can cause the groundwater to become contaminated and the water table to drop. Irrigation can also lead to the salinisation of soils as the irrigation water, which contains dissolved salts from the underlying geology, evaporates leaving the salts behind.

Education

Informing farmers of the long-term consequences of their farming techniques as well as methods to minimise these consequences, remains one of the most effective ways of combating land degradation. This is an integral part of the work done by the Soil Conservation Service (SCS). Representatives from the SCS offer farmers advice and support in employing farming methods that will ensure the long-term sustainability of agriculture in the Great Plains. This type of work is an on-going process. Memories can be short and if individual farmers have no experience of the damage and hardship which can result from inappropriate farming methods, they may 'cut corners' in order to increase profits.

CASE STUDY
The Sahel

In arid climates land degradation can lead to **desertification**. Fundamentally, desertification means land turning to desert. Desertification threatens approximately one-third of the Earth's surface, affecting the lives of 1.2 billion people. As Figure 2.2 shows, the regions most susceptible to the process of desertification are those bordering the world's hot deserts, for example, the Sahel, which borders the Sahara.

It is estimated that 6 million hectares of productive land are lost each year to desertification. The Food and Agricultural Organisation of the United Nations estimated in 1998 that, if world soils continue to degrade at their present rate, in around 20 years the world will be unable to produce enough food to feed the rapidly increasing global population.

Figure 2.42 Location of the Sahel region

A combination of physical, climatic and human factors have made the Sahel region of Africa particularly susceptible to desertification.

Location

The Sahel runs parallel to the southern edge of the Sahara Desert from the west coast to the east coast of Africa. The average width of the Sahel is 500 km. Its proximity to the Sahara is one of the principal reasons it is prone to desertification (Figure 2.42).

Climate

The principal features of the Sahel climate are shown in Figure 2.43.

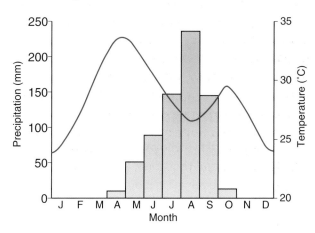

Figure 2.43 Climate graph for Sokoto in the Sahel

◆ Total annual rainfall is low, ranging between 200 mm in the northern Sahel to 500 mm in the southern Sahel.
◆ Rainfall is unreliable and highly variable.
◆ Rainfall often occurs as an intense tropical downpour, creating high potential erosivity.
◆ There is a distinct dry season – that is, rainfall only occurs during certain months. There are high rates of evaporation due to high

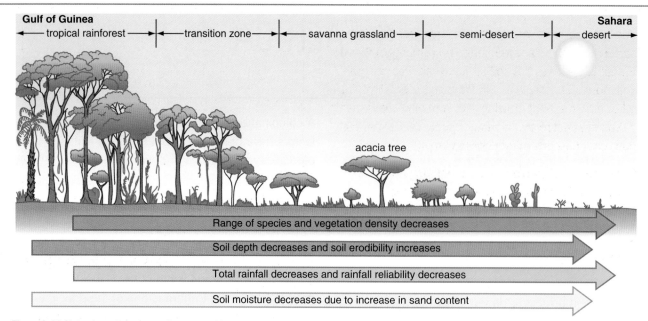

Gulf of Guinea ←tropical rainforest→ ←transition zone→ ←savanna grassland→ ←semi-desert→ **Sahara** ←desert→

acacia tree

Range of species and vegetation density decreases

Soil depth decreases and soil erodibility increases

Total rainfall decreases and rainfall reliability decreases

Soil moisture decreases due to increase in sand content

Figure 2.44 Natural vegetation zones in western Africa

temperatures which means much of the precipitation is lost.

The rainfall pattern in the Sahel region is a result of the **Intertropical Convergence Zone (ITCZ)** which migrates across the region each year. During the dry season, Tropical Continental air is dominant along with dry **harmattan** winds blowing south across the Sahara.

Natural environment

Due to the variability of the rainfall across the Sahel, the soils and natural vegetation also vary. In southern regions where the rainfall is more reliable, well developed, fertile soils are found with a greater variety of flora and fauna. Moving north, vegetation becomes less dense and the soils have a low organic and clay content (Figure 2.44). This means they are more susceptible to the processes of erosion.

Human activity

Traditionally there have been two main types of subsistence agriculture practised in the Sahel.

Arable farming

A wide variety of crops, for example, barley, sorghum and vegetables, are grown together with drought-resistant species like millet. Trees are retained and/or planted amongst the crops. These provide protection from wind erosion by acting as a windbreak, binding the soil together and by shading the ground (which reduces evaporation). The trees will also provide a source of fuel for cooking and fruit/nuts for the farmer. Farmers leave sections of their land fallow for up to twenty years to allow it to rejuvenate.

It is important to grow a variety of crops in order to avoid the impact of monoculture, or relying on only

QUESTIONS

1 What is desertification and why does it affect the Sahel?
2 a Using Figures 2.43 and 2.44, describe the features of climate in the Sahel that make it susceptible to desertification.

b The climate graph does not tell us anything about rainfall reliability. Why is this information crucial in being able to understand why the Sahel is affected by desertification?

one crop, which exhausts the soil by removing the same nutrients year after year. The removal of these nutrients destroys the soil structure and leaves the soil more susceptible to wind and water erosion. The other advantage of growing a variety of crops is that if a crop is affected by disease or pests the farmer still has other sources of food for the family.

In this way the arable farmer is able to provide a mixed diet for the family while retaining the productivity of the soil by varying the crops grown.

Pastoral farming

Livestock rearing of cattle, goats and sheep is also practised in the Sahel. The majority of these pastoral farmers are **nomadic**, moving from one area to the next following the rains, to find fresh pasture for their animals. The success of this type of farming depends on:

◆ avoiding overgrazing an area, which leaves the soil vulnerable to erosion, as the vegetation no longer offers any protection
◆ finding sources of water for animals to drink (wells, rivers, lakes)
◆ the ITCZ providing sufficient rainfall for vegetation to grow in the area and so maintaining herd sizes at a level that is sustainable.

In many regions of the Sahel, the size (not the quality) of an individual's herd is a measure of his status within a tribe. It is a hard but highly adapted way of life. Some tribes such as the Samburu of northern Kenya are able to survive for long periods on what their cattle can provide – for example, they drink a mixture of the blood and milk, eat the meat and wash in the urine.

3 Explain why the traditional agricultural techniques practised in the Sahel have been successful for many centuries and have not resulted in the long term damage of the Sahel's ecosystem.

Figure 2.45 Rainfall variation in Burkina Faso

Causes of rural land degradation in the Sahel

The causes of rural land degradation in the Sahel cannot be attributed to any one factor, but to the interaction of a wide range of both physical and human variables.

Physical factors

Droughts are now occurring in the Sahel with greater frequency than has been experienced previously. Figure 2.45 shows the rainfall variation in Burkina Faso. It is estimated that between 1920 and 1995 Burkina Faso's water table dropped by 20 m. This has major implications for the growth of vegetation and the formation of soils. As the vegetation

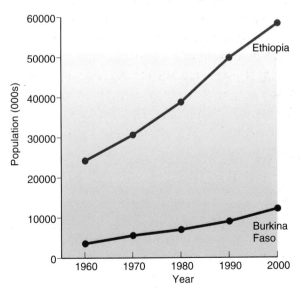

Figure 2.46 Population growth in the Sahel

struggles to survive in the drought conditions, the soil becomes exposed and the topsoil (containing the vast majority of the fertility) is blown away by the wind or washed away by flash floods caused by tropical downpours. The unpredictability of the Sahel climate is illustrated by the fact that Burkina Faso's cotton crop was destroyed by flash floods in 1994; and in 1996–97, 67 000 tonnes of emergency food aid were required by Burkina Faso to deal with food shortages caused by drought.

Human factors

Improved medical care resulting in reduced death rates, coupled with high birth rates, mean that rapid population growth is occurring in the Sahel, as shown in Figure 2.46. With more mouths to feed, greater pressure is placed on the finite resources available to produce more food. In countries such as Burkina Faso, Chad, Mali and Mauritania, populations are increasing at a rate of 2.5% per year (while at the same time food production increases at only 1% per year). This has contributed to a number of factors which have caused the desertification of the Sahel.

◆ **Overcultivation** – Arable farmers are forced to increase yields from their land. As a result, the fallow periods allowed for the soil to regenerate are reduced and the soil becomes depleted of its nutrients. With rainfall in the region increasingly

Figure 2.48 Forest land turned to desert by overgrazing and deforestation

unreliable, crops often fail and the soil is degraded. Rising populations have also forced farmers to increase the amount of land utilised for arable farming. This has led to the cultivation of marginal areas, for example, on the fringes of the Sahara. These areas are not a viable option in terms of sustained crop production and, as a result, cultivating them accelerates the process of desertification (Figure 2.47).

◆ **Overgrazing** – The rise in population has brought about an increase in the size and number of the herds grazing in the Sahel. This places greater pressure on grazing areas, particularly around the wells, boreholes, lakes and rivers where the animals are taken to drink. The concentration of herds within these areas results in the vegetation being stripped down to the roots, leaving soil exposed. In addition, compaction of the soil by trampling reduces the infiltration capacity of the soil and increases runoff (and therefore water erosion). As a result farmers are forced to graze increasingly marginal areas. Land degradation follows rapidly, allowing the desert to expand.

◆ **Deforestation** – Rising populations place an ever increasing burden on woodland which constitutes the main source of domestic fuel in the Sahel. Since 1900, 90% of forests have been cleared from the Ethiopian Highlands. Deforestation removes the binding effect of tree roots, the protection afforded by the foliage and the ability of the trees to reduce wind speeds (Figure 2.48). The shortage of **fuelwood** means

Figure 2.47 Millet field during the dry season in Mali

that more and more families are having to resort to animal dung and crop residues for their domestic energy requirements. The dung and crop residues would normally be used as fertiliser so the soil is further degraded as it is deprived of essential nutrients. In the long term, this will affect crop yields.

◆ **Urbanisation** – In many parts of the Sahel there has been a rapid growth in urban populations as people move from rural areas in search of jobs, a higher standard of living and better education opportunities. The population of Niger's capital, Niamey, rose from 207 000 in 1980 to just under 1 million in the year 2000. The growth and development of small settlements in rural areas has also increased, as schools have been set up, as in Korr in northern Kenya. All these settlements create a growing demand for firewood, so large areas of forest are cut down and the wood sold in the settlements. Much of this fuel is sold as charcoal because it is lighter, not as bulky and therefore cheaper than wood to transport. The wood is transformed into charcoal by burning it very slowly in earthen pits. This process is extremely inefficient and half the energy of the wood is lost during conversion. It is estimated that 1.3 billion people worldwide are consuming fuelwood resources at a rate that is not sustainable.

◆ **Growth of cash crops** – During the twentieth century some farmers in the Sahel moved away from traditional techniques and began growing cash crops such as cotton and rice. This resulted in monoculture, which rendered the soil infertile. This was often accompanied by inappropriate farming techniques such as flood irrigation. In Niger, for example, water is pumped out of the River Niger to irrigate fields of rice. The flood irrigation can lead to salinisation of the soil and as a result it becomes degraded.

QUESTIONS

4 'Urbanisation is the main cause of land degradation in the Sahel.' Do you agree or disagree with this statement? Justify your answer.

5 The promotion of cash crop agriculture presents a conflict of interests for countries in the Sahel. Discuss the arguments for and against increasing the cropland area given over to cash crops.

Roots of irrigated crops reject most salts, absorbing nearly pure water. To flush the salts away from the roots, extra water must be used - but that can raise the water table

In a dry climate, low rainfall generally keeps the water table depressed, and vegetation is sparse

Without proper drainage, the water table continues rising, and the water evaporates from upper layers of soil. Since only pure water evaporates, salt concentrations in the soil increase and can eventually poison crops

Figure 2.49 How irrigation can cause salinisation of the soil

Consequences of land degradation

Physical impacts

The increasing frequency of drought periods, the farming of marginal areas along the northern boundary of the Sahel and deforestation have led to the southward expansion of the Sahara Desert. In the last 50 years, 65 million hectares of the Sahel have turned to desert and in Sudan the Sahara has advanced by 100 km in 17 years. Desertification is now estimated to be spreading at a rate of 1.5 million hectares per year in the Sahel.

◆ **Rills and gullies** – The exposure of soil leaves it open to the intense tropical downpours of the ITCZ. As a result, overland flow removes topsoil and cuts into the land, forming rills and gullies. This is also a problem where soil has been compacted by the trampling of animals.

◆ **Loss of topsoil** – Strong winds, such as the harmattan, erode the topsoil from land that has become degraded due to overcultivation, overgrazing and deforestation. During the drought of the 1970s, three times the normal amount of topsoil was blown away. Some of this was deposited in Barbados nearly 5000 km away.

◆ **Salinisation** – Inappropriate farming technology and the growth in cash crop farming has led to an increase in flood irrigation, causing salts to accumulate in the soil (Figure 2.49). If irrigated soils are not drained adequately, salts will build up in the root zones of crops. This can kill the crops and contaminate the soil. To reverse the process and rejuvenate the soil is very expensive.

Social and economic impacts

◆ **Malnutrition and starvation** – The failure of crops year after year leads to starvation and death, as in Sudan and Ethiopia in the mid-1980s. Malnutrition may also be attributed to the reduction in the range of crops grown in order to concentrate on cash crops.

◆ **Migration** – The degradation of farmland and the increased pressures created by rising populations have forced people to migrate away from their home areas in the Sahel. This has led to the loss of traditional farming techniques and in some cases, for example, in northern Niger, a demographically imbalanced rural population. This can occur when younger men move away in search of an alternative income, leaving the women, children and the elderly at home.

◆ **Dependence on external support** – Where the effects of land degradation and drought have been most acute, people have come to rely on food and other aid sent from other countries, for example, following the Band Aid campaign of 1983. This can lead to **over-dependency** on external help.

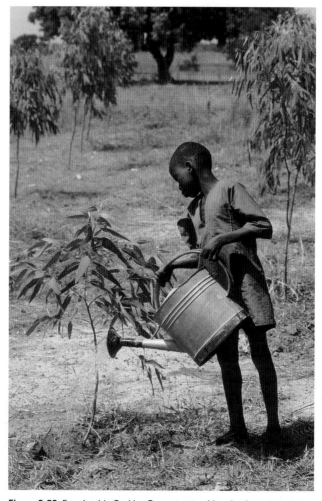

Figure 2.50 Farmland in Burkina Faso recovered by planting new trees

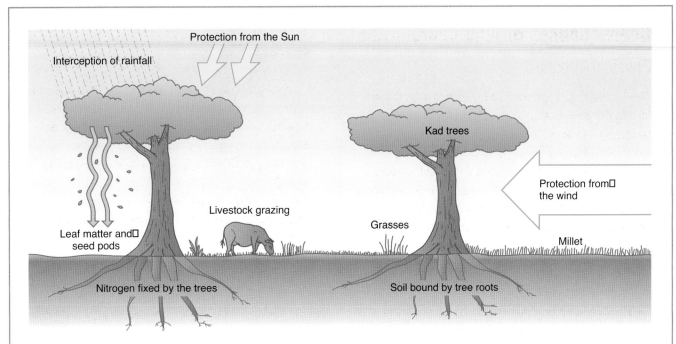

Figure 2.51 Agroforestry techniques

◆ **Access to education** – In many of the Sahelian countries, such as Burkina Faso, education and health care must be paid for and therefore the loss of income brought about by the failure of crops and herds may mean students have their schooling interrupted and individuals go untreated when ill.

◆ **Loss of income to farmers** – this is the the most obvious economic impact of rural land degradation in the Sahel. In addition, as the yields of cash crops decrease because of the exhaustion of the soil, the quantities of commodities, such as cotton, being exported will fall and countries will lose a valuable source of foreign currency.

Solutions to land degradation

Afforestation

Planting new trees helps to bind the soil, and also provides shade, windbreaks (reducing wind erosion), nutrients and fuel for families (Figure 2.50). Depending on the species planted the trees may also provide nuts and fruit for humans and animals. One important element of any afforestation

programme is to ensure the supply of fuelwood is sustainable. This has led to the practice of **agroforestry** (Figure 2.51) where farmers utilise the benefits offered by the trees for their crops or for grazing their animals. The trees help shade crops and animals from the harsh sunlight and provide an additional source of nutrients for the crops (in the form of decaying leaf matter).

In parts of West Africa, the Kad tree, a variety of Acacia, has been used in an attempt to restore soil fertility and productivity (Figure 2.51). The advantages of the Kad are that:

Figure 2.52 Agroforestry in Mali

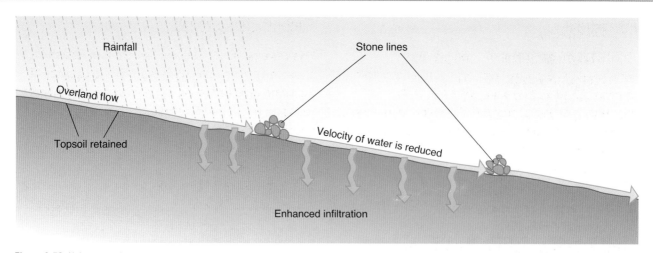

Figure 2.53 Using stone lines to prevent soil erosion

- ◆ it retains its leaves in the dry season, so providing shade for the soil
- ◆ it fixes nitrogen in the soil, thus raising fertility
- ◆ its seeds and pods provide a rich source of protein for cattle and goats.

Stone lines

The construction of low-lying **stone lines** (or diguettes) along the contours of the land is a very effective way of reducing soil erosion and the loss of water by overland flow (Figure 2.53). In addition it can be completed as part of a community project with families helping each other at very little cost (assuming that a source of stones is available). Communities in Mali, Sudan, Burkina Faso (Figure 2.54) and Niger, with the support of aid organisations such as Oxfam and Tear Fund, have used this method very effectively. In some cases crop yields have been increased by as much as 50%.

Managed grazing areas

Research and tests have shown that grasses are able to re-establish themselves very quickly given an undisturbed fallow period in which to recover (1–2 years). Therefore if areas are fenced off from grazing animals on a rotational basis it is possible to sustain the herds without long-term damage to the soil. The main disadvantage of this method is the cost of fencing and the management of the grazing

Figure 2.54 Stone lines in Burkina Faso

areas, for example, around the settlement of Korr in northern Kenya.

Reduced herd sizes

Reducing the numbers of cattle, sheep and goats and focusing on the quality rather than quantity of the herds, decreases grazing pressure on the land. It also reduces the impact of soil compaction which causes increased overland flow and reduced infiltration. This is a very culturally sensitive issue as traditionally the size of a herd is often a measure of an individual's status within his tribe and therefore herders may be resistant to change.

Education

Teaching farmers about the causes and consequences of rural land degradation can help change attitudes. An example of this is the importance of the appropriate irrigation techniques which avoid the salinisation of the soil, for example, drip irrigation. Unfortunately these alternatives are often more expensive.

Fuel-efficient stoves

The development of stoves like the **jiko**, which is cheap and uses approximately half the quantity of fuelwood required by an open fire, can play a part in reducing rural land degradation. If these stoves are used, the demand for fuelwood is reduced and trees will be retained for longer. This technique can be particularly effective in areas where the cost of alternative fuels, such as kerosene, are beyond the means of local people.

QUESTIONS

6 Discuss why trees are seen by many as the key to reducing and reversing rural land degradation in the Sahel.

7 a It is now widely recognised that any techniques used to combat land degradation in the Sahel must be 'appropriate'. What does the term 'appropriate' mean in the context of combating rural land degradation in the Sahel?

b To what extent can the solutions outlined in this section be viewed as being appropriate? Outline the main advantages and disadvantages of each of them.

CASE STUDY
Livestock project in Burkina Faso

Background

Burkina Faso has a population of 9.4 million, and this is increasing at a rate of 2.5% per annum. The location of Burkina Faso is shown in Figure 2.55. Life expectancy is 48 years and approximately 85% of the population live in rural areas. Adult literacy is low (8.4% for females and 27.9% for males).

Burkina Faso faces a number of fundamental problems which include:

◆ desertification and land degradation caused by overuse of natural resources and a series of drought years

◆ the challenge of feeding a rapidly increasing population while relying heavily on basic farming methods which have not changed in centuries

◆ government policies and growing consumerism which have led to greater pressure to grow cash crops

◆ a heavy national debt problem

◆ AIDS, which is a major threat and is already affecting the balance of the population and reducing debt repayment.

This livestock project is based in the Tangaye, 140 km north of Ouagadougou (Figure 2.56), an area farmed by the Mossi tribe. Tangaye is an agro-pastoral zone where subsistence farming is the main livelihood. Land degradation and inadequate rainfall have made it increasingly difficult for families here to produce sufficient food. One of the main consequences of these factors has been a significant rise in rural–urban migration and the

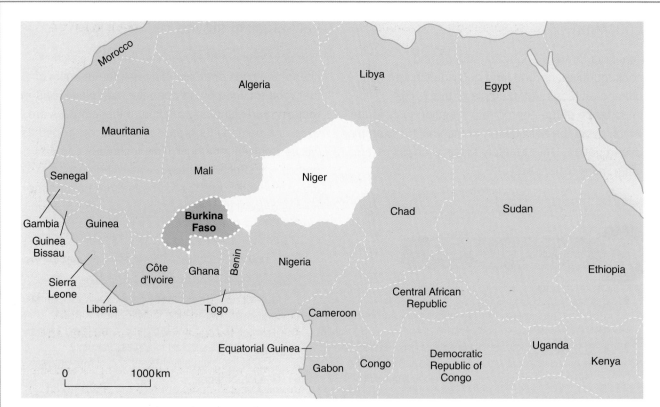

Figure 2.55 Location of Burkina Faso and Niger

Figure 2.56 Location of the Tear Fund project in Burkina Faso

subsequent breaking up of rural communities. The project is supported by the British charity Tear Fund.

Aims of the project

The initial aims of the project were to provide improved living standards and greater food security for the families in Tangaye. This was to be achieved through provision and training in improved agricultural techniques. The objectives were to:

◆ increase food production and diversify crops
◆ improve soil fertility through the use of animal manure
◆ reduce the burden of work on families, especially women, and increase their productivity through the use of appropriate technology.

Methods used:

◆ Education and training was provided for farmers in the selection and care of their oxen.
◆ Manure-producing cowsheds were established. These enable the animals to be well looked after and at the same time produce a reliable source of manure which can be used to improve soil fertility. The cowsheds were to be paid for by the farmers, although other equipment such as

feeding troughs, spades and scythes were funded by Tear Fund. The manure is produced in a pit (4 m diameter and 4 m deep). During the dry season, water is added to the pit to ensure the manure rots fully. This is important because if the manure has not rotted sufficiently, it can actually stunt plant growth.

◆ One animal (cow) was bought for each family.

◆ Agricultural equipment was provided, in particular ploughs. By using cattle to do the heavy work on the land, farmers are able to cultivate a larger area more efficiently. In order to cultivate an area of 100 hectares, 20 cattle and 20 ploughs are needed.

◆ Veterinary care is provided. The cattle involved in the project were vaccinated in order to ensure they were healthy enough to carry out the work necessary.

◆ Stone lines were constructed along contours to retain water, decrease runoff and increase infiltration.

◆ A community approach was encouraged. The farmers were arranged into groups to encourage mutual support. In addition a system was set up whereby after three years families included in the project agree to sell their cattle and buy three younger animals. Of the three purchased, two

are retained and the third is given to another family.

◆ Some reforestation was also carried out.

Outcomes

◆ Families are able to cultivate between two and three times the area of land they could by hand.

◆ Food production has increased. It is estimated that crop yields are around 25% higher when manure is used.

◆ Other families in the area are now copying the techniques introduced. The number of cattle has doubled in the last five years.

◆ Standards of health have improved, particularly amongst women as they are able to use the animals for transport and are saving time and energy.

◆ The main problem facing the project is enabling it to become self-sufficient and therefore no longer reliant on outside agencies for funding.

◆ There have also been a small number of deaths amongst the cattle, which obviously inhibits the effectiveness of the project and highlights the need for adequate veterinary care to be available and accessible to the local people.

CASE STUDY
Integrated development project in Niger

Background

This project was established to help the Tuareg ethnic group in Niger (Figure 2.55). The Tuareg are semi-nomadic pastoralists whose methods have changed little over the centuries. In recent years they have experienced increasing hardship through the loss of large numbers of livestock. This has been as a result of drought, rising population, desertification and the encroachment of settled agriculture.

The focus of the project was in an area north-east of Abalak (Figure 2.57). It is an area of semi-arid grasslands, with a total annual rainfall of around 250 mm. Three fossilised valleys dissect the region, each with an intermittent river during the rainy season. The area was affected by severe droughts in 1993–94 and 1997–98. Most of the soils are loose, sandy and retain very little moisture. However, the soils on the flood plains of the valleys have a higher clay content and are agriculturally viable. Large areas of natural forest have died out since 1984.

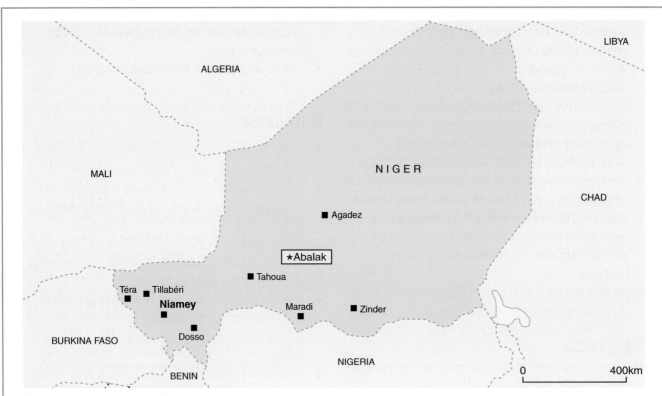

Figure 2.57 Location of the Tear Fund project in Niger

This has left the soil exposed; erosion has followed and any natural regeneration of the forest is impeded by animals. Herbaceous plant species found in the area provide excellent fodder but local people have reported that at least twelve species have disappeared since 1984. The principal products of the Tuareg's animal husbandry are milk, cheese, leather, skins and meat.

Aims

The main objective of the project was the establishment of **fixation points** – fixed areas where the Tuareg can spend most of the year. The aim of these fixation points is to 'provide the population with the ability to improve, manage and use the available natural resources, and survive droughts and famines'. Secondary to this is the hope that fixation points will provide the opportunity to develop and integrate improved education and health services, while still retaining local culture and tradition. It is important to emphasise that this is not **sedentarisation** or the permanent settlement of people in one place. The Tuareg still migrate, but the fixation point develops a central area where they already spend the majority of the year.

Methods

- ◆ Stone lines were constructed to improve water retention and infiltration.
- ◆ Five wells have been created and five repaired.
- ◆ The cutting and selling of deadwood provides additional income but it must be sustainable.
- ◆ Traditional migration during the rainy season is encouraged.
- ◆ Training is provided in the production of mud bricks and woodless construction techniques. Communal house building events are organised. The aim is to reduce the Tuareg's reliance on wood for this purpose.
- ◆ Better-quality breeds of livestock have been introduced. Herds are traditionally large, inefficient and unproductive. The introduction of alternatives means herd sizes can be reduced.
- ◆ Economic diversification is encouraged – the people are being educated about alternative sources of income as herding alone cannot

support the increasing population. This is a long-term goal of the project.

- Pasture management – this includes the fencing off of high-quality pasture, the rotation of livestock and regeneration of pasture, so that everyone benefits.
- Afforestation – tree nurseries have been established and from this a tree planting programme has been developed.
- Provision of tools and seeds – the aim here is to encourage the Tuareg to cultivate crops in addition to rearing their herds. This in turn provides them with an alternative source of food and reduces their reliance on and vulnerability to market forces.

Outcomes

- It is estimated the project has benefited around 2500 people directly and about the same number indirectly.
- The stone lines have led to the regeneration of some herbaceous plants, the establishment of pasture and the cultivation of crops such as sorghum, millet and cowpeas. The construction

of one stone line can regenerate at least 20 hectares of land.

- The Tuareg's food supply is now more diverse and reliable.
- Afforestation is now taking place as a result of the seedling nurseries.
- A much stronger community approach to the problems facing the Tuareg is developing. This is a particularly significant development given the fiercely independent nature of the Tuareg.
- The traditional attitudes towards herd sizes are being changed as the people see the benefits to be gained from having a smaller, better-quality herd.
- The additional income being generated by overall diversification is enabling the Tuareg to send their children to school.
- The rate of change is very slow due to the strength of the Tuareg's traditions and culture.
- Conflicts have been experienced between the Tuareg and other ethnic groups, for example, the Fulani, who are moving into the area because of the widespread problems of rural land degradation in Niger.
- The project is still reliant on funding from outside agencies, such as Tear Fund.

QUESTIONS

With reference to the last two case studies:

8 Outline why each project was necessary – it may be helpful to do this using a spider diagram.
9 Are the techniques used in each project appropriate to the area they serve? Justify your answer.
10 Draw up a table of the main benefits and drawbacks of each project.
11 Using the relevant information from the last two case studies, draw up two revision tables using the headings given below, and complete them – one for the Sahel and one for the Great Plains. Remember to record as many facts, figures and examples as you can.

Causes of rural land degradation	Consequences of rural land degradation	Solutions and their effectiveness

12 'Climatic factors are the main cause of rural land degradation.' Do you agree with this statement? Support your answer using examples from *either* North Africa *or* North America.
13 Describe in detail the impact that land degradation has had on the **physical landscape** of the Sahel.
14 Explain, giving examples, why many of the techniques used to combat land degradation in North America would be considered inappropriate in the case of the Sahel.
15 'Water erosion has a more devastating impact in terms of land degradation than wind erosion.' Discuss the validity of this statement, using examples from the Sahel and North America to support your answer.

Key terms and concepts

- aeolian erosion (p. 91)
- afforestation (p. 103)
- agroforestry (p. 103)
- arable farming (p. 98)
- biological action (p. 70)
- bison ranching (p. 95)
- boom or centre-pivot irrigation (p. 96)
- cell grazing (p. 95)
- chemical erosion (p. 70)
- chernozem (p. 89)
- contour ploughing (p. 85)
- crop rotation (p. 86)
- deforestation (p. 100)
- deposition (p. 72)
- desertification (p. 97)
- detachment (p. 72)
- double cropping (p. 86)
- Dust Bowl (p. 93)
- erodibility (p. 75)
- erosion (p. 69)
- erosivity (p. 75)
- fallow period (p. 95)
- fixation points (p. 108)
- fuelwood (p. 100)
- gully erosion (p. 72)
- harmattan (p. 98)
- infiltration rate (p. 77)
- intercropping (p. 86)
- Intertropical Convergence Zone (ITCZ) (p. 98)
- jiko (p. 105)
- laser levelling (p. 94)
- malnutrition (p. 102)
- minimum tillage (p. 94)
- monoculture (p. 82)
- mulching (p. 95)
- nomadic (p. 99)
- overcultivation (p. 100)
- overgrazing (p. 71)
- pastoral farming (p. 99)
- physical erosion (p. 70)
- rainsplash (p. 71)
- rill erosion (p. 72)
- runoff (p. 86)
- Sahel (p. 97)
- salinisation (p. 71)
- saltation (p. 75)
- sedentarisation (p. 108)
- sheet wash and sheet flow (p. 71)
- shelter belt (p. 93)
- Soil Conservation Service (SCS) (p. 95)
- stone lines (p. 104)
- strip cropping (p. 85)
- surface creep (p. 75)
- surface runoff (p. 76)
- suspension (p. 74)
- sweep plough (p. 94)
- tillage (p. 94)
- transportation (p. 72)

SUMMARY

Having worked through this chapter, you should now know:

★ the nature and processes of soil erosion and land degradation

★ the effects of climate and its variability on these processes within North America and North Africa

★ that ecosystems are modified by human activity

★ the physical, social and economic impacts of soil erosion and land degradation

★ specific examples of soil conservation and land management strategies and their effectiveness through case studies from North America and Africa.

Section 3

URBAN CHANGE AND ITS MANAGEMENT

Links to the core

The movement of population from rural to urban areas has now become the world's biggest migration pattern. As a consequence an increasingly large proportion of the world's population is now classified as urban. This section will build on the ideas and concepts introduced as part of the Urban Geography section of the core unit: urban systems, urban functions and structures, and urban change. It examines the phenomenon of urbanisation and focuses particularly on the problems of managing the changes which face growing urban areas. Relevant aspects of other parts of the Human core, such as population geography and industrial geography, have been fully integrated into the case studies. Two cities will be studied in depth: Toronto in Canada, exemplifying managing change in a rapidly growing city in a more economically developed country (MEDC); and Karachi in Pakistan, an expanding city in a less economically developed country (LEDC).

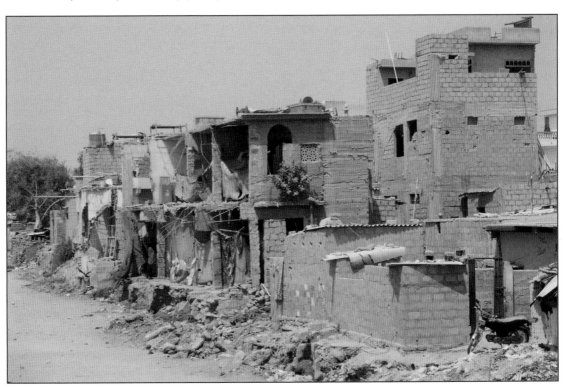

Population growth in Canada

Canada is the world's second largest country in area (9 970 610 km²) and is divided into a number of provinces and territories (Figure 3.1). The 2001 Census counted 31 021 300 people in Canada – just over half the population of the UK. The Canadian population doubled in the 50 years after 1951 and although the rate of growth slowed between 1991 and 1996, the five-year period to 2001 saw a return to a high rate of growth (Table 3.1), slowing once more in the period to 2005. None the less, Canada's population continues to grow at the second fastest rate of all the G8 industrialised nations (Table 3.2).

The distribution of Canada's population is also significant. Over 80% of the Canadian population is concentrated into a narrow strip a few hundred kilometres wide along the length of the US–Canada border. As Figure 3.2 shows, despite its vast area, this concentration of population indicates a high degree of urbanisation.

Urbanisation in Canada

Most of the rapid expansion of Canadian cities has taken place since 1950. This contrasts sharply with the experience of many UK cities which have experienced major losses of population due to the process of counter-urbanisation during the same period.

For example, the population of Greater London fell by 200 000 between 1951 and 1961, and there was a

Figure 3.1 Canadian provinces and territories

1 Lake Superior
2 Lake Michigan
3 Lake Huron
4 Lake Erie
5 Lake Ontario

Table 3.1 Population growth in Canada since 1951

	Total population	Population increase	Growth rate %
1951	14 009 429		
1956	16 080 791	2 071 362	14.8
1961	18 238 247	2 157 456	13.4
1966	20 014 880	1 776 633	9.7
1971	21 568 311	1 553 431	7.8
1976	22 992 604	1 424 293	6.6
1981	24 343 181	1 350 577	5.9
1986	25 309 331	966 150	4.0
1991	27 296 859	1 987 528	7.9
1996	28 846 761	1 549 902	5.7
2001	31 021 300	2 174 539	7.5
2005	32 270 500	1 249 200	4.0

Source: World Population 1996, United Nations, Population Division

further loss of 540 000 between 1961 and 1971. Liverpool, Manchester and Newcastle each lost around 18% of their respective populations during the 1960s. The 1981–91 period in the UK revealed that although population decentralisation continued, there were some interesting exceptions. The major metropolitan centres of Manchester, Liverpool, Sheffield, Newcastle, Birmingham, Leeds and Glasgow showed continually high rates of out-migration, but Inner London experienced considerable population growth, as have several non-metropolitan urban areas.

The three largest Canadian cities have experienced very rapid growth rates post-1950 and have at least doubled their populations. Toronto quadrupled its population during this time and in 1976 overtook Montreal as Canada's largest city. During the same period, Vancouver established itself as the third largest city in Canada. The 2001 Census revealed that the population increase in the 20 largest **Census Metropolitan Areas (CMAs)** was 1.2 million, representing 60% of the total Canadian population growth (Table 3.3). A CMA consists of a very large urban core together with adjacent urban and rural areas that are economically and socially related to the core. The urban core has a minimum population of 100 000. The fastest growing cities during the period 1991–2001 were Calgary (+26.2%), Vancouver (+24.0%), Oshawa (+23.5%), Windsor (+22.2%) and Toronto (+20.3%). More than a third of Canada's population lived in the four most populous metropolitan areas: Toronto, Montreal, Vancouver and Ottawa. The 2001 Census also confirmed a trend, seen for a number of years, showing the

Table 3.2 Population growth rates in G8 nations, 1994–2004

	Population as of 1 July		Annual growth rate (%)
	1994	2004	1994 to 2004
Canada	28 999 006	31 974 363	0.98
United States	263 125 821	293 655 404	1.10
Germany	81 338 100	82 531 700	0.15
Federation of Russia	148 446 239	143 974 059	−0.31
France	57 565 000	59 900 700	0.40
Italy	56 843 400	57 888 200	0.18
Japan	125 265 000	127 687 000	0.19
United Kingdom	57 862 100	59 834 900	0.34

Sources: Statistics Canada, US Census Bureau, Eurostat, UK Office of National Statistics, Statistics Bureau of Japan, Russian Federal State Statistics Service

Figure 3.2 Distribution of population in Canada

distribution of Canada's population gradually shifting from east to west. Significant growth in the two major cities of Alberta – Calgary and Edmonton – can be accounted for by the development of provincial oil reserves and consequent industrial development along with expansion of public services. In Ontario, the population increased by 10.6%, of which 60% was accounted for by international migration and the remainder by natural increase and migrations from other Canadian provinces. About 80% of Ontario's growth occurred in its CMAs and half of the entire population growth was within the Toronto CMA. In 1996 the Toronto CMA became the first in Canada to be populated with over 4 million people.

There are three main causes of rapid urbanisation in Canada. First, the cities have attracted substantial numbers of migrants from rural areas. Tables 3.3 and 3.4 show a clear trend where the larger urban areas have experienced growth, whilst those more

dependent on primary industries such as fishing in Newfoundland (St John's) and mining in Ontario (Sudbury) show small population losses. Significantly, both Newfoundland, hit by the collapse of its fisheries in the early 1990s, and Saskatchewan, heavily dependent on prairie farming, showed their first recorded losses of population as migrants left for other parts of the country.

Second, the largest Canadian cities have traditionally been the destination of overseas immigrants settling in Canada. Although the source of immigrants to Canada has changed, during the 1950s, 1960s and 1970s immigrants came mainly from Europe, whereas today they are mainly from Asia; the trend for them to settle, initially at least, in the main urban centres has continued. International migration accounted for over half of the population growth recorded in Toronto between 1996 and 2001. These immigrants, often young married couples, are responsible for the third cause of population growth

Table 3.3 The 20 largest urban areas in Canada, 1991/2001 (population figures in thousands)

Rank in 2001	Metropolitan Area	1991	2001	% change 1991–2001
1	Toronto (Ont.)	3893	4683	20.3
2	Montreal (Que.)	3127	3426	9.6
3	Vancouver (B.C.)	1603	1987	24.0
4	Ottawa–Hull (Que.)	921	1064	15.5
5	Calgary (Alta.)	754	951	26.2
6	Edmonton (Alta.)	840	938	11.6
7	Quebec (Que.)	646	683	5.7
8	Winnipeg (Man.)	652	671	3.0
9	Hamilton (Ont.)	600	662	10.4
10	London (Ont.)	382	432	13.2
11	Kitchener (Ont.)	356	414	16.4
12	St Catharines–Niagara (Ont.)	365	377	3.3
13	Halifax (N.S.)	321	359	11.9
14	Victoria (B.C.)	288	312	8.3
15	Windsor (Ont.)	252	308	22.2
16	Oshawa (Ont.)	240	296	23.5
17	Saskatoon (Sask.)	210	226	7.6
18	Regina (Sask.)	192	193	0.4
19	St John's (Nfld)	174	173	−0.5
20	Greater Sudbury (Ont.)	158	156	−1.5

by generating higher birth rates for the cities than for other parts of Canada.

Urban trends in Canada

Although each Canadian city has responded to the demands of population growth in different ways, it is possible to identify a number of common trends:

◆ the outward physical spread of each city in response to demands for low-density suburban housing on the outskirts

◆ pressures on agricultural land immediately outside the city limits for change of use to housing, light industry and office parks

◆ heavy pressures on the network of highways and public transport systems as a result of the movement of population, particularly at peak hours

◆ major changes in land use within the cities in response to changing requirements of commerce and industry.

Table 3.4 Population growth from 2001 to 2005 in the Canadian provinces

	2001	2002	2003	2004	2005	2001–05 absolute change (thousands)	% change
	population (thousands)						
Canada	**31 021.3**	**31 372.6**	**31 669.2**	**31 974.4**	**32 270.5**	**1 249.2**	**4.03**
Newfoundland and Labrador	522.0	519.4	518.5	517.3	516.0	−6.0	−1.1
Prince Edward Island	136.7	136.9	137.3	137.9	138.1	1.4	1.02
Nova Scotia	932.4	934.5	936.3	937.5	937.9	5.5	0.58
New Brunswick	749.9	750.3	751.2	752.1	752.0	2.1	2.80
Quebec	7397.0	7445.7	7494.0	7547.7	7598.1	201.1	2.71
Ontario	11 897.6	12 102.0	12 259.6	12 407.3	12 541.4	643.8	5.41
Manitoba	1151.3	1155.6	1161.6	1170.2	1177.6	26.3	2.28
Saskatchewan	1000.1	995.9	994.5	994.3	994.1	−6.0	−0.61
Alberta	3056.7	3116.3	3159.6	3204.8	3256.8	200.1	6.54
British Columbia	4078.4	4115.4	4154.6	4201.9	4254.5	176.1	4.31
Yukon Territory	30.1	30.1	30.6	30.9	31.0	1.0	2.99
Northwest Territories	40.8	41.5	42.2	42.9	43.0	2.2	5.39
Nunavut	28.1	28.7	29.2	29.7	30.0	1.9	6.76

Note: Population as of 1 July. Source: Statistics Canada

Figure 3.3 Canadianville

Also, in terms of physical appearance, there is an increasing similarity between Canadian cities. This is best represented by a study of their profiles, which have assumed a pyramid or inverted cone shape (Figure 3.3). The apex is located at the heart of the **Central Business District (CBD)** which is usually the site of the tallest building or skyscraper. With increasing distance away from this point, building height decreases. This classic profile can be explained in terms of land values and rents. Economic and competitive market forces have created the central skylines of Canadian cities. Despite changes in central areas of European cities in recent years, it is true to say that their skylines are often dominated by churches, cathedrals and castles and it is certainly not possible with any certainty to point from a distance and identify their peak land values. In Canadian cities this can be done with monotonous ease.

QUESTIONS

1 Explain why 'Canada is more urbanised . . . than many Western European countries'.
2 Outline the differences in the pattern of urban growth in Canada between 1951 and 1981 compared with that which took place in most Western European countries during the same period.
3 On an outline map of Canada locate and mark the 20 largest cities.
4 What were the reasons for the rapid growth of Canadian cities between 1951 and 2001?
5 Explain why population growth from 1996 to 2001 was greatest in Ontario, Alberta and British Columbia.
6 Describe and explain the significance of the Canadianville model.
7 Explain why competition for land is greatest in the Central Business Districts of Canadian cities.

CASE STUDY
Toronto

Situation, site and growth

The Huron Indians called the present site of Toronto 'the carrying place' because it was the centre of the area's fur trade. The area was strategically important in controlling major trade routes and first the French, and later the British, built a series of forts in the area (Figure 3.4). In 1788 the British purchased control of the area from the Mississauga Indians, paying 146 barrels of blankets, axes and cloth. The first lieutenant-governor of Upper Canada, John Graves Simcoe, had his doubts about the value of the purchase and commented that 'the city's site was better calculated for a frog pond or a beaver meadow than for the residence of human beings'. Today, the land is worth many billions of dollars and is home to over 4 million people.

Originally named York, the population reached 1000 by 1818, and in 1834 the town became a city and was renamed Toronto. The city grew quickly during the nineteenth century mainly due to the advantages of its site and situation. The **site** – that is the actual piece of ground on which the city has developed – was chosen because of its large, natural harbour on Lake Ontario sheltered by a peninsula which later became the Toronto Islands (Figure 3.5). The **spit**, formed from clay, sand and gravel, was separated from the shoreline in 1858 following a major storm and the hooked shape of the islands creates a natural harbour.

Many features of the site of Toronto show the influence of glaciation. At the end of the ice age Lake Ontario was much bigger than it is today and was called Lake Iroquois. Meltwater was trapped behind ice which blocked the natural exit channel down the St Lawrence Valley, so instead water flowed southwards and created the Hudson–Mohawk routeway to New York. The old Lake Iroquois shoreline in Toronto is marked by a steep **bluff** or slope between 20 and 25 m in height (Figure 3.6). The construction problems caused by the bluff meant that much of Toronto's early growth took place

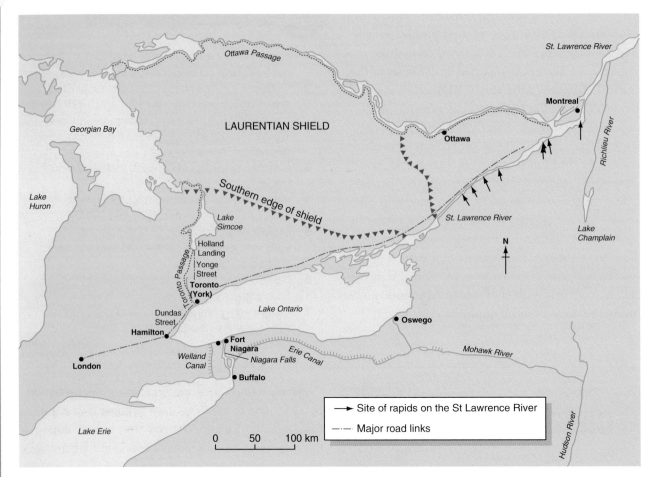

Figure 3.4 Location of Toronto

east–west rather than towards the north. South of the bluff the Iroquois plain allowed easy construction and a ready supply of building materials from the sands and gravels deposited along it. Extensive gravel bars were deposited where the Humber and Don Rivers entered the former Lake Iroquois (Figure 3.5).

The site of the city is greatly influenced by the valleys of a number of rivers and creeks which flow into Lake Ontario (Figure 3.6). The most significant of these are the River Humber, River Don, Highland Creek and the Rouge River. These rivers and their tributaries have cut deep **ravines** into the **till plain** deposited by glaciers, and have restricted urban development to the land in between them. The ravines themselves are unsuitable for building because of their low-lying nature and also because much of their surface material is unconsolidated. In

October 1954, Hurricane Hazel hit the city and 36 people were killed where housing had been built on the Humber flood plain. Bridges have had to be built to allow east–west routeways across the modern city, although the valleys themselves have been used for transportation corridors (Figure 3.7) and some form extensive parklands (see Figure 3.23).

The original old town was laid out on the lake plain. As transportation networks developed during the second part of the nineteenth century, Toronto began to expand rapidly. Industrial growth was stimulated by the railways and the development of the port. The construction of the Welland and Erie Canals improved the city's links with New York via the Mohawk and Hudson Rivers. The coming of the railways reinforced Toronto's position as the hub of transportation with improved links with the rest of Canada and the north-eastern USA. This early

Figure 3.5 Physical site of Toronto

Figure 3.6 Topography of Toronto

Figure 3.7 The Don Valley expressway

growth took place to the south of the core area. As industry developed and more European immigrants arrived in search of jobs, housing was largely constructed to the east and west of the original site of the city. These houses were mainly three-storey semi-detached buildings on 5 m wide plots. They were built at such a high density that by 1945 the city had one of the highest population densities in North America. By the early 1950s, with continuing migration from other parts of Canada and increased immigration, the city was expanding rapidly and the urban structure developing to include neighbouring municipalities.

QUESTIONS

8 Outline both the advantages and disadvantages of the site of Toronto for urban growth.
9 Using an atlas and Figure 3.5, explain why Montreal was better located for the development of international trade than Toronto, before the mid-nineteenth century.
10 Explain the advantages to Toronto of the construction of the Welland and Erie Canals.
11 In what ways was continued European immigration vital to the development of Toronto in the late nineteenth and early twentieth centuries?

Population distribution and growth

As well as being Canada's largest city, Toronto is now the eighth largest urban centre in North America. In 2001 the Toronto Census Metropolitan Area was home to 4 682 897 people, a growth rate of over 20% since 1991. The Greater Toronto Area (GTA) (Figure 3.8) includes the City of Toronto itself and the regions of Halton, Peel, York and Durham. The population of the city area was 2.481 million in 2001 accounting for 48.8% of the GTA population. Most of the recent growth has taken place in the outlying regions where the rate of population increase is three times that of the city itself. Toronto is much more compact than the GTA regions: in 2001 Toronto's average population density was 3939 people per km^2, compared with a combined average

Figure 3.8 The Greater Toronto Area (GTA)

of 400 for the regions. Densities in Toronto's urban core reach as high as 6740 people per km^2. The main reason for the lower density in the outer area is the desire of many people to purchase single-family homes on large lots and land is much cheaper further away from Toronto. There is, in fact, almost no open land remaining within the city boundaries.

Indications are that growth will continue to be greater in the regions and it is estimated that by 2011, 55% of the GTA's population will live outside the city. The populations of Peel and York are expected to double in the next 20 years. Much of this growth is due to the availability of jobs, with almost 50 000 in government alone in Toronto.

Toronto's population, like that of the country as a whole, is ageing. By and large Toronto is ageing and older, whilst the surrounding regions are younger. Toronto has 49% of the working age and 59% of the senior GTA populations. Many senior citizens remain in the houses they purchased after the Second World War and into the 1970s. Others prefer to live in the city because of transport accessibility and equally accessible social and human services. The 'greying' of Toronto's population is reinforced by the tendency for many young families to move to larger houses in the regions around the city, leaving the older population living in the city itself. Evidence of this trend is found in the distribution of children: the regions have 52% of pre-school and 55% of school age children. Despite the outward movement of population towards the suburbs, the population of the City of Toronto itself is expected to grow slightly due to two identifiable trends:

◆ new housing areas are being constructed on old **brownfield** industrial sites such as the railway lands near the lake
◆ young adults are moving into older areas of the city where they renovate older houses (**gentrification**).

Ethnic groups and inequalities

Toronto continues to be one of Canada's primary immigrant reception areas, and as a result the city is ethnically very diverse. The city's big magnets are

Figure 3.9 Major immigrant concentrations in Toronto

the availability of rental housing and the presence of its existing immigrant communities. Between 1996 and 2001, 280 650 people emigrated to Toronto. Providing low-cost housing for such large numbers is becoming increasingly difficult. Federal and provincial housing projects have been drastically cut and the number of new rental units in the city cannot meet existing or predicted demand. As a result, homelessness in Toronto has reached unprecedented levels, evidenced by the increasing numbers of people living on the streets, using public buildings as shelters and relying on food banks and other charitable services. It is estimated that on any given night, over 3200 people use Toronto's hostels.

As indicated by Figure 3.9 the longest-established immigrants, such as those from Italy and the UK, tend to have the highest proportions already moved from the central to outlying parts of the city. In the 1950s, 75% of the City of Toronto's population was of British origin and even by 1981 the figure was 48%; by the 2001 census it had fallen to 7.5%.

Initially the UK was the main source of Toronto's immigrants, followed by Italy. More recently this has switched to China, India, Pakistan, the Philippines and Sri Lanka. There is a tendency for certain areas of the city to house concentrations of immigrants from one country (Figure 3.9), but overall Toronto is a cosmopolitan city, with immigrants gradually establishing themselves and moving out from the city centre to better housing in the surrounding suburbs. Perhaps one of the best examples of this is the Chinese community which has traditionally been centred in Toronto's Chinatown located on the edge of the CBD.

'Toronto boasts about its downtown Chinatown. It is one of the multi-cultural neighbourhoods the city celebrates with special features, such as street signs in the resident's mother tongue. The area has the busy feel of a multi-ethnic street carnival. Signs and storefronts blaze with a patchwork of mandarin red, emerald green and imperial gold. Scores of restaurants venture far beyond the fried rice, chicken balls and chow mein served across North America, offering foods from every region of China. Dozens of

shops display herbal remedies such as ginseng and lotus root, as well as clothing, hand-hammered woks and other housewares, furniture, ornate jewellery and Chinese books and records. The shelves and bins of grocery stores are packed with specialities such as bok choy and Chinese cabbage – now widely grown on southern Ontario farms – along with sweet, yellow Chinese pears, hot bean sauce, sticky brown rice paste and chicken feet.

Chinese people can find almost any service, or conduct almost any transaction, in their own languages. Large numbers of Chinese lawyers, doctors, herbalists and acupuncturists have set up practices. Streets, banks, police stations and most other services are identified by both English words and Chinese characters. Four theatres screen Chinese movies and there are branches of several Hong Kong banks. At night the streetscape of shop signs and store windows create a colourful tapestry of lights. The activity lasts hours past midnight: many of the restaurants stay open until four or five in the morning.'

'Farewell to Chinatown', Peter Gorrie, Canadian Geographic

In recent years successful Chinese-Canadians have been moving uptown to Scarborough, some 25 km north of the CBD. Located amongst one of Toronto's sprawling post-war suburbs, this Chinatown is far bigger than the one in the city centre and it is mainly the young and affluent Chinese who are making the move, leaving behind an ageing population in the traditional Chinatown. The City of Toronto's Chinese population increased from 5000 in 1951 to over 350 000 in 2000, with many new immigrants now going straight to the suburbs if they can afford to. Most of those moving into the downtown quarter are from mainland China and other parts of south-east Asia. They have little education and few skills and many are illegal immigrants; a few are criminals who contribute to the area's high crime rate and there are now fears of a racist backlash. The Chinese community have been discriminated against in Canada's past. In 1885 the federal government imposed a $50 head tax on each Chinese immigrant,

12 Describe and explain the growth and change in the distribution of Toronto's population since 1950.

13 Outline the problems for city planners which have resulted from the large increase in population in the suburban areas of the city, and in the surrounding regions.

14 Which groups of people are moving into the inner city area and why?

15 Using Figure 3.9, outline the evidence which suggests that the Italians were amongst the first 'non-UK' immigrants in Toronto.

16 From those shown on the maps in Figure 3.9, which nationality are the latest group of immigrants? Support your answer with map evidence.

17 Describe and explain the distribution of the Chinese population in Toronto.

18 What visual evidence might you observe on a visit to the city to support the claim that Toronto is a cosmopolitan city?

which was increased to $500 by 1893. Between 1923 and 1947, legislation barred Chinese immigration into Canada and even today the Chinese are under-represented in most elected positions. The recent transition to the suburbs may herald the start of a new era for Toronto's growing Chinese community.

Managing urban change in Toronto

Following the Second World War, Toronto grew very rapidly and a new form of local government structure was required to manage and plan for the changes that were taking place in its urban structure. Although the population of the City of Toronto itself was not increasing, the populations of the surrounding suburbs, each of which had its own

municipal government, were exploding. With the high density of inner city housing, and the rapid growth of car ownership in the post-war years, demand for housing and schools in the suburbs was increasing at alarming rates. It was necessary to provide an overview of the entire region for planning purposes and a new system of local government was established which has since been modified several times in response to the continually changing geography of the Toronto area.

◆ In 1953 the City of Toronto and its twelve suburbs were amalgamated under a regional government called Metropolitan Toronto. Responsibility for strategic planning was taken over by the new Metro authority but local services such as education and many community services were retained by local councils.

◆ In 1967 the thirteen municipalities were reduced to six: the cities of Toronto, Etobicoke, North York, Scarborough, York and the Borough of East York (Figure 3.10).

◆ In 1998 these were amalgamated into one authority called the City of Toronto. The electoral wards are shown in Figure 3.11. The creation of the newly defined city in 1998 was recognition that Toronto is the centre of a broader Greater Toronto Area within which the regions of Halton, Peel, Durham and York and the City of Toronto itself are mutually dependent.

Figure 3.10 City of Toronto

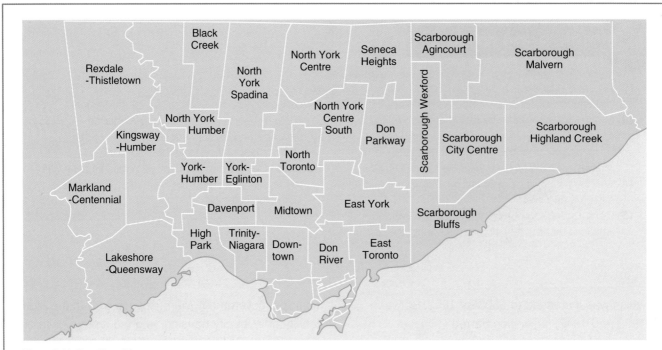

Figure 3.11 New City of Toronto

Rouge
River
Valley

B₁

DA

B₂

TIA

C₃

C₁

C₂

A

T

N

Metropolitan major centres

B₁ North York

B₂ Scarborough

Metropolitan intermediate centres

C₁ Etobicoke

C₂ Yonge-St. Clair

C₃ Yonge-Eglinton

C₄ Kennedy-Eglinton

TIA Toronto international airport

T Toronto Island airport

DA Downsview airport

0 1 2 3 4 5 km
Scale

☐ Pre 1945 housing

☐ Post 1945 housing

☐ Industrial, railway and port areas

☐ Open space (public & private)

A CBD

Figure 3.12 Land use in Toronto

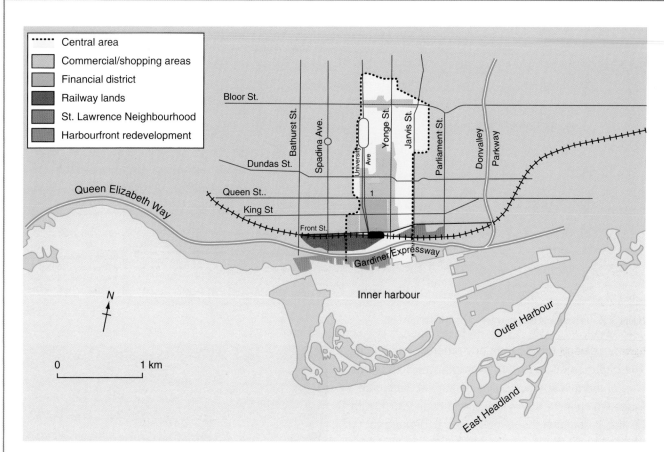

Figure 3.13 Toronto City centre and Harbour area

Managing land use change

As the city has grown and developed, its neighbourhoods have tended to take on specific characteristics, and to be used for specific purposes. Through an analysis of the land use pattern which has emerged (Figure 3.12) the changes and current problems facing the city can be identified.

Central Business District

The location of Toronto's CBD is shown in Figure 3.12. Toronto is Canada's leading financial centre and the greatest part of that activity takes place within the CBD. Toronto's CBD extends south to the railway lands, north to Bloor Street, west to University Avenue and east to Jarvis Street (Figure 3.13).

As in any large city, the CBD contains the greatest concentration of tall buildings (Figure 3.14). The

Figure 3.14 CBD of Toronto

nearer a building is to the CBD, the taller it is likely to be because that is where land is most expensive. The central area functions as a primary location for government, corporate head offices, financial institutions, retailing and tourist attractions. Toronto

Figure 3.15 Underground pathways and malls in Toronto's CBD

Above ground view
1 Union Station
2 Royal York
3 Royal Bank
4 55 University
5 Commercial Union
6 Aetna Centre
7 Toronto Dominion Tower
8 Royal Trust Tower
9 Sun Life
10 Commerce Court
11 Montreal Trust
12 First Canadian Place
13 Hilton Hotel
14 Sheraton Centre
15 Simpsons
16 City Hall
17 Bell Telephone
18 Eaton Centre
19 Atrium on Bay

Subway stations
A Osgoode   D King
B St. Andrew   E Queen
C Union   F Dundas

Existing underground path
Future underground path
Subway line

has more head offices than any other Canadian city. The high-rise office towers average over 30 storeys. One of the tallest office buildings is First Canadian Place which is 72 storeys high and rises 290 m with 58 lifts. It contains thousands of legal, brokerage, banking and other business offices. On an average working day the complex is occupied by approximately 30 000 people whose total annual income exceeds a billion dollars. These people need to work in close proximity for meetings and other business activities, although increasing use of information technology is making business clustering of this type less necessary, and many corporations are now looking at cheaper out-of-town office locations such as that at North York. In fact, in recent years office development in the CBD has slowed considerably due to high land prices and traffic and parking problems.

Toronto's City and Metro Halls and the Ontario Legislative Assembly buildings are in the CBD as well as most provincial government offices, foreign consulates and overseas trade missions.

The CBD is the hub of many transport routes: all of the subway lines converge on the CBD which is also the centre of a huge regional transportation system run by the Government of Ontario; the main bus terminal and Union Station are located towards the

Figure 3.16 The SkyDome and CN Tower

south of the CBD. Most of the office buildings are linked by underground pedestrian routes called PATH (Figure 3.15). Although these are very useful during the cold winter months, the system was not planned coherently and routes between the 1100 stores connected by the system can be confusing.

Despite the increase of the suburban shopping malls, many people prefer the noise and activity of Yonge Street (Figure 3.13) for shopping which has thousands of stores, and also the Eaton Centre, a CBD shopping mall, easily reached by subway link from all parts of the city. The centre has 300 shops, cinemas and restaurants on four climate-controlled levels which span many city blocks. This downtown core retailing area ranks third in retail sales in North America, after Manhattan and Chicago. In 2003, 17 100 retail businesses provided over 11% of the city's employment.

Finally, Toronto's CBD is also a tourist and entertainment centre boasting attractions such as the CN Tower, theatres, concert halls and sports arenas, for example, the SkyDome Stadium with its unique retracting roof (Figure 3.16).

Retailing

About 5% of land use in the Greater Toronto Area is devoted to commerce (shops and offices). Despite moves towards **decentralisation** of both retail and commerce from the CBD, the greatest concentration of both remains in the city centre. Outside the city centre it is possible to identify three different levels of provision:

◆ **Strip-commercial** – small stores, offices and petrol stations lining both sides of a major street in a ribbon pattern. Arteries such as Yonge Street and Bloor Street (Figure 3.13) are examples.

◆ **Shopping plazas** – consisting of around 10–20 shops of a low-order type serving neighbourhood areas of a few thousand people. They usually include a food store, variety store, bakery, drug store and a few other small outlets.

◆ **Regional shopping malls** – there are currently nine of these huge enclosed centres in Toronto, strategically located near major highways with massive free parking lots. Each is anchored by one or more major Canadian department store along with 100 to 200 chain and speciality stores.

Suburban business districts: the emergence of edge cities

Although the population of Toronto is gradually moving away from the older parts of the city, most employment in offices and retailing remains in the CBD. The central area retains around 500 000 jobs, half of which are held by people commuting each day from outside the City of Toronto. Overall land use figures indicate a slight increase in office use but hide the massive expansion that has taken place mainly through **vertical development** in recent years due to high land values. Since 1971 the total

Figure 3.17 Metropolitan centres and corridors in Toronto

amount of office space in central Toronto has more than tripled. This increase, together with the population movement to the suburbs, has resulted in greatly increased pressure on the highway and public transportation system. As with retailing, there has been a trend towards the development of **suburban business districts**, notably in North York, Scarborough and Etobicoke (Figure 3.17). All are located on subway lines and that at Scarborough is at the end of the new light rapid transit system; it combines the Education Centre with several high-rise office blocks and a huge shopping mall. By 2000 the complex was employing 40 000 people.

The development of certain growth points on the edges of existing cities, which may consist of business parks, shopping complexes and new housing areas, has been identified as a major new urban phenomenon – the **edge city**. In the City of Toronto, it is possible to identify a number of these, for example, North York, Scarborough and Mississauga. Although they are at different stages of development, these growth points have certain common characteristics:

◆ more than 1.5 million m² of leasable office space
◆ more than 180 000 m² of leasable retail space
◆ more jobs than bedrooms
◆ most growth has taken place within the last 30 years
◆ each is perceived by the public as one place.

Significantly, these edge cities now mirror the downtown functions of the CBD. In North York, cheaper land and rents have attracted several corporations to move their headquarters or other operations to the new development. More than 3.2 million m² of space is provided with the creation of 100 000 new jobs in shops, hotels and offices. A total of 60 000 people are expected to live in the new high-rise flats and **condominiums**. The scheme has been criticised on several fronts: for lacking human scale, for being sterile and uninviting, and for the fact that none of the new housing will be aimed at low-income families.

QUESTIONS

19 Outline how and why urban growth has affected local government in the Toronto area since 1950.
20 Summarise the main land uses within Toronto's CBD.
21 Identify four different levels of retail provision in Toronto and comment on their functions.
22 Explain the reasons behind the emergence of suburban business districts.
23 Identify examples of edge cities in Toronto's urban structure and outline their main characteristics.

Industry

With more than 4 million residents and 2.5 million jobs, this is the most populated and largest economic region in Canada. Toronto has the largest labour force of any city in Canada. One in five employees works in manufacturing or warehousing, and the Toronto region has over 7000 manufacturing companies accounting for 21% of the Canadian total. Traditionally, this area has been the industrial heartland of the country. The late twentieth century saw considerable industrial change and restructuring but car-making, telecommunications, pharmaceuticals, chemicals and plastics and the aerospace industries continue to expand and invest in the Toronto area. The importance of this mix of traditional and emerging industries lies in their potential for future growth. In particular the area's positive economic outlook is enhanced by:

◆ easy access to large markets in the USA
◆ a centrally located international airport
◆ a highly integrated motorway system
◆ a well-educated labour force
◆ training, research and development at local universities.

When considering the industrial geography of the Toronto area, it is important to recognise the economic forces and the historical factors which have resulted in the existing pattern of industrial activity. Any analysis of this distribution requires an understanding of the many influences which provide certain locations with competitive advantages.

Before 1850 manufacturing was largely confined to small workshops within the city area. Small markets for manufactured goods and limited accessibility to those markets were the main factors responsible for the small scale of production. After 1850 the emergence of the railway led to a significant improvement in accessibility to markets, and towards the end of the nineteenth century Toronto emerged as a centre of industrial activity. This continued into the twentieth century, with the advent of mass production technologies and assembly lines which contributed to single-storey modern factories. Increasingly larger scales of production in these factories began to have significant negative impacts on the areas immediately around them, and there was a need to separate industrial activities from other land uses. Additionally, with increased dependence on road and air transport, new industrial parks characterised by large, low-density, single-purpose buildings began to emerge near to the international airport and the major highways around Toronto (Figure 3.12).

Distribution of industry

Between 10 and 15% of Toronto's land is used for industry. The distribution of manufacturing industry (Figure 3.12) reflects the continuing importance of historical factors and inertia in the inner city area, and the outstanding importance of road transport in the newer areas of the outer suburbs.

Western and eastern edges of the CBD – These areas became important in the late nineteenth and early twentieth centuries, having access to both labour and markets. They are now in decline due to outdated premises and the difficulties of expansion. The main industry still located in this area is clothing which employs 10 000 workers (often female immigrant workers of Greek, Italian and Portuguese origin). The area acts as a 'conveyor belt', with small companies each specialising in individual stages of making a garment. Wholesalers and distributors are

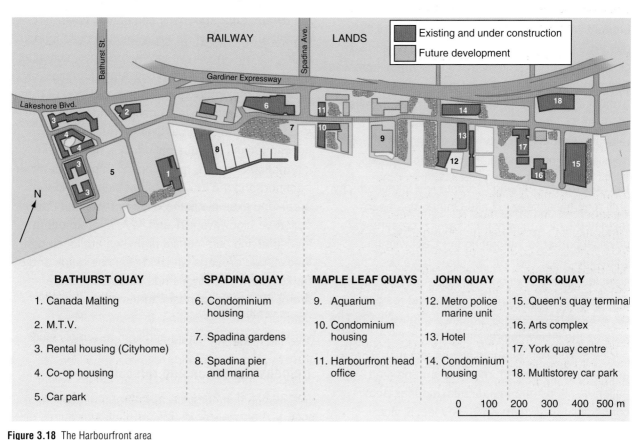

Figure 3.18 The Harbourfront area

BATHURST QUAY	SPADINA QUAY	MAPLE LEAF QUAYS	JOHN QUAY	YORK QUAY
1. Canada Malting	6. Condominium housing	9. Aquarium	12. Metro police marine unit	15. Queen's quay terminal
2. M.T.V.	7. Spadina gardens	10. Condominium housing	13. Hotel	16. Arts complex
3. Rental housing (Cityhome)	8. Spadina pier and marina	11. Harbourfront head office	14. Condominium housing	17. York quay centre
4. Co-op housing				18. Multistorey car park
5. Car park				

Figure 3.19 Harbourfront, Toronto

also concentrated in this area and high land prices can be borne by having firms on different floors of multi-storey buildings.

Don/Harbour Area – This is the oldest industrial area, based around the harbour where many industries requiring imported raw materials located to cut costs. Food processing (flour milling, sugar refining, whisky distilling) used raw materials brought in by water, and used the lake for waste disposal. The main railway yards were adjacent, and the area also developed meat-processing factories as cattle were brought into the city by rail from other parts of the country. The area was flat, easy to build on and was close to the workers' houses at a time when there was little personal transport. Today this area is undergoing a major change: no space for expansion, congested roads and no parking are forcing industries to relocate. As they do so, the area is gradually being transformed into housing and recreation space.

Harbourfront – Although Toronto is a lakeshore city, until recently access to Lake Ontario has been restricted by the railway tracks which presented a physical barrier to integrating the CBD with the lakeshore area. The last few years have seen the complete transformation of the old harbourfront area as new commercial, entertainment and residential developers have seen its potential (Figure 3.18). The area by the lake was previously an industrial and dockland zone of warehouses and factories. These are now rapidly being turned into art and craft galleries, shops and restaurants, marinas and condominiums (Figure 3.19). The last is vital

because it is an important part of the plan to have people living full-time in the area (although so many high-rise apartment buildings have now been constructed that some people argue these have created a new barrier between the city and its waterfront).

Between Harbourfront and the CBD other new developments are under construction in a large area formerly used for railway yards for assembling freight trains. The area is adjacent to the CBD and therefore a prime development site. The CN Tower (at 533 m the world's largest free-standing structure) and the SkyDome, with its retractable roof, are both located here (Figure 3.16). A programme of further development began in 1993 and recently new office buildings, shopping and entertainment areas and apartments for 15 000 people have been completed. It is the largest redevelopment project in Toronto's history and is scheduled for completion by 2010.

Industrial parks in the suburbs

Land and taxes are cheaper in the regions outside the central area and many companies have relocated to sites which allow them to build modern space-consuming single-storey buildings, suited to modern automated technology. Access to highways is another major factor, as is the avoidance of older, congested sites. The labour supply has also relocated to the suburbs. One of the major concentrations of industrial parks is near the international airport, reflecting the importance of air transport and the focus of major highways in the area. One example is the Skyway Industrial Park, built near highways 401 and 427, where modern industrial and commercial premises are located in landscaped grounds with wide service roads and car parks. The type of industries attracted to the area are of the 'light-secondary' tertiary/quaternary category: data processing, insurance offices, warehousing, packaging and light assembly work.

Deindustrialisation and relocation

In planning industrial strategies for the twenty-first century, planning authorities in Toronto have identified the relatively high percentage of vacant industrial floor space in the central area of the city as **deindustrialisation** of these areas. City planners are now concerned about the loss of manufacturing jobs to the outer edges of the built-up area such as North York, Scarborough and Etobicoke and beyond into the regions. Between 1950 and 2000 the number of industrial jobs in the central area halved. Although increased office jobs have offset this loss to some degree, the economic base of the central area has been narrowed, and job opportunities reduced. Massey-Ferguson was one of the largest inner-city firms, but it contracted from 2000 workers in 1976 to 500 workers in 1985. Despite reducing the size of the plant, overheads remained high and traffic congestion reduced efficiency. The entire site has now been closed and has been developed for housing and some light industry. Massey-Ferguson has relocated to a cheaper site at Brantford between Toronto and London, Ontario. The Toronto Economic Development Corporation was created to maintain a diversified economy in the centre, but despite incentives for prospective employers, high land prices, property taxes and traffic congestion continue to force expanding manufacturers to relocate in the regions that border the city. This reveals a clear pattern of **industrial suburbanisation**.

The effect of this deindustrialisation on the ground is very clear to see. Demolition of older industrial buildings and the resulting vacant land and possible soil contamination discourage potential developers from locating in these areas and encourage further decline. Once-thriving concentrations of employment within the city are becoming areas of derelict land. In contrast, as more and more land on the fringes of the built-up area is converted to industrial use, farmland is lost to developers and the boundaries of the urbanised area expand ever outwards. Perhaps the most significant implication of the changing distribution of industrial activity in Toronto is its effect on the efficient use of the urban infrastructure. The abandonment of former industrial land in the central area leaves existing services such as utilities and transport unused. At the same time, the level of

demand for these services in high-growth areas on the urban fringe increases.

QUESTIONS

24 What advantages does Toronto have as an industrial location?
25 Explain what you understand by the term 'industrial suburbanisation'.
26 What problems have been caused by industrial suburbanisation in Toronto for:
   a inner city areas
   b peripheral areas of the city?
27 Identify and explain why some inner city industries have not relocated to the suburbs.
28 Outline the redevelopment of the Harbourfront area of Toronto.

Housing

Over 80% of land in Toronto is zoned for residential use. As shown in Figure 3.12, most of the land used for housing is in the suburbs. However, Toronto is unusual amongst major North American cities in having a considerable number of residential buildings in or near the downtown area. Most of these are high-rise apartment buildings and obtaining low-cost housing is very difficult. At the end of the twentieth century new condominiums were still being built in the CBD. 'Condos' are simply apartments that are owned by their residents. This unusual concentration of residents in central areas

Figure 3.20 Detached housing in suburban Toronto

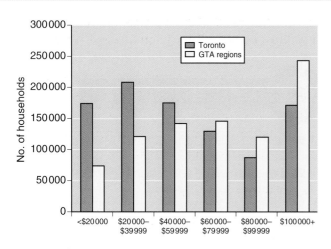

Figure 3.21 Households by income groups

means that Toronto's CBD is busy almost 24 hours a day, helping to create a safer downtown. This is reflected in the title of the city's structure plan, **'The Liveable Metropolis'**.

Within Toronto the choice of residence is dictated largely by income. The cost of housing is increasing faster than in any other Canadian city. For most residents, the preferred choice of accommodation is a detached house in the suburbs (Figure 3.20), but for many this is becoming less possible and **income polarisation** is increasing. This means that most households in the city are concentrated at the lower end of the income spectrum, whilst those in the GTA regions are polarised at the high end. In 2001 the City of Toronto had 72% of the area's lower income households, whilst the surrounding regions had 59% of all high income households (Figure 3.21). Such polarisation between the **core and periphery** is common and perhaps even more pronounced than in other cities in North America. Such an imbalance creates problems for Toronto's tax base and the provision of social and human services. It is important to realise that just what makes up the 'core' and the 'periphery' is subject to change through time. Older housing predominates in the downtown area, with post-war housing in the outer city areas of Etobicoke, North York and Scarborough. These areas were the periphery in the middle of the twentieth century and grew rapidly after 1945 as young families moved out to suburban

neighbourhoods where identical houses were built over vast areas. Scarborough in particular became the subject of many jokes as 'Scarberia' or the 'bedroom city' dormitory suburb.

During the 1970s new spatial patterns in housing emerged with huge increases in population in the fringe regions around the city (which were the new peripheral areas as suburbanisation had spread even further from downtown). Between 1971 and 1981, Etobicoke, North York and Scarborough lost population as some of their residents moved even further from the CBD to the newly developing outer suburbs.

Today in the city itself there is a concentration of rental apartments, condominiums and assisted housing. Low to moderate income families live here, together with single people, immigrants and senior citizens. In the outer regions, detached housing predominates and this region is predominantly populated by young, upper income families.

It is almost exclusively young families buying a home in the outer regions, where their preferred house type (a detached house) is more affordable, who leave the city. During the 1990s each year over 100 000 people left the city to live in the surrounding regions. Net migration from the city accounted for half of the GTA regions' growth during this decade.

Immigration accounts for a significant flow into the city each year. Over 70% of immigrants settle in the city because of the availability of rented accommodation. In addition, there are return flows of migrants from the GTA regions; these are mostly students leaving the family suburban home and renting flats and rooms in the inner city near the universities.

Since the flow of population out of the city is overwhelmingly the young, there is a knock-on effect on the structure of the population that remains in the city, which is only marginally tempered by immigration. By 2011 over half of the city's home-owners will be over 55, and a similar ageing pattern is evident amongst the tenant population, many of whom are unable to afford the move to the higher-cost housing in the suburbs.

Inner city redevelopment: St Lawrence

A number of major inner city renewal projects were undertaken by the city towards the end of the twentieth century. They were part of a process initiated by the city planners called **reurbanisation** which is intended to reverse the seemingly unstoppable outward urban sprawl by undertaking **urban renewal** projects in areas where the **urban fabric** of services is already in place.

The St Lawrence area (Figure 3.13) is located just to the east of the CBD and was the largest inner city housing redevelopment scheme undertaken during the twentieth century. The area is spatially distinct, with the office and retailing core on one side, an old industrial area to the east, and the railway to the south. Working-class housing was developed in the St Lawrence area during the nineteenth century along with storage and distribution facilities for industry. By the mid-1970s, housing quality had declined and the industrial functions had largely moved out to the suburbs, leaving large areas derelict. The city council purchased any land it did not already own and established a plan for the area based on integrating the new development with the surrounding area and avoiding becoming a public housing ghetto for poorer income families:

◆ traditional street patterns were largely retained with shops and medium-rise apartment blocks
◆ several architects were employed to avoid the 'Scarberia' factor
◆ a mix of private and subsidised housing was provided.

As shown in Figure 3.22, the redevelopment included a new school and a young people's theatre, and retained the famous St Lawrence Market, a popular local indoor market much used by CBD office workers.

The development of the St Lawrence area was seen as a rejection of the policies of city redevelopment, which are still popular in many European cities, where high-rise flats were thought to provide the answer to inner city housing problems. It also avoided the pitfalls of inner city redevelopments in the USA, notably in Detroit and Chicago, which became no-go terror zones. The scheme has been completed and now houses some 10 000 people with a mix of household types, incomes and lifestyles. The success of the St Lawrence neighbourhood has proved that Toronto can deliver on its promise of a more liveable city.

Open spaces

From the Toronto Structure Plan:

'The Metropolitan open space system shall continue to be based primarily on the major river valleys and the waterfront ...'

Despite the rapid growth of the city since 1950, Toronto has managed to retain many unique and interesting natural areas. Of 7200 hectares of parkland, currently 3273 hectares are controlled by the Metro Parks department and include managed areas such as horticultural and zoological gardens and golf courses, and wilderness areas deliberately left as natural as possible (Figure 3.23). The city's many natural features are mainly found along the Lake Ontario waterfront and along the river valley and stream corridors. These natural areas include an abundance of unique geological and biological features which support equally unique and diverse plant and animal life and associated habitats. Toronto is a remarkably green city with its natural areas offering a range of outdoor recreational opportunities.

The Valley Lands

The Humber, Don and Rouge Rivers, together with several tributary creeks, have cut deeply incised valleys through the easily eroded sediments of the former lake (Iroquois) bed. These slopes are relatively unstable and building regulations prohibit construction on them. In many cases they have been afforested to help with **slope stabilisation** and to reduce runoff. This has also enhanced their scenic attractiveness as they cross the cityscape. The shape of the open spaces follows the valleys (Figure

Figure 3.22 Redevelopment of the St Lawrence area

PHASE C

PHASE A

PHASE B

0   100   200 km

1. Private housing
2. Private condominium units & commercial
3. Mixed private/public housing
4. St. Lawrence Market
5. Gross machinery site

6. Housing
7. Market site, housing & commercial
8. Private condominium units
9. Private commercial development
10. Young people's theatre
11. Cityhome
12. Transformer site

13. Cityhome
14. Private housing
15. Co-op housing
16. Private commercial development
17. Co-op housing
18. School site

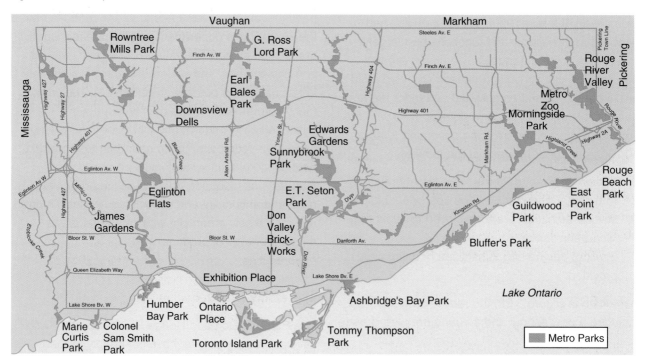

Figure 3.23 Open spaces in Toronto

3.23) and they are ideal for walkways and cycle tracks.

The large block of open space in the east (Figure 3.23) is the Rouge river valley which has been designated as Canada's largest urban park (4250 hectares). The land has remained undisturbed because it was purchased in the mid-1970s as greenbelt land between the city and a proposed new airport. Once the airport proposal was dropped, the fight over the Rouge valley began between developers and conservationists. The developers rented space on billboards next to the expressways and commuter routes near the valley and put up adverts claiming, 'If it weren't for a few deer, you'd be home by now!' The conservationists pointed out the effects of development near the mouth of the river where manicured park lawns have replaced the tall grasses and other natural bird habitats. Further west, the Don and Humber valleys are even worse. Little is left of the Don's marshes and the Humber's are stagnant with toxic industrial effluent. In contrast, the Rouge valley remains mostly unspoilt with oak, cedar, hemlock and maple woodlands and open farmland. Even with park protection, pressure from the surrounding urban area still exists, for example, to use the valley for landfill sites, golf courses and even cemeteries! The valley represents one of Toronto's few remaining intact ecosystems and will need the protection given by its recent designation as an urban park.

The Waterfront

Toronto's lake frontage gives it great potential for a range of watersports and shoreline recreational activities. One project already very successful in this respect is the Harbourfront development discussed on page 132. Other schemes include the Toronto Islands with walks, beaches and a traditional Canadian village; and Ontario Place, a 40 hectare recreation complex including a marina, an arena for rock concerts and a six-storey Imax cinema. An important part of the city's plan for this area is to create a lakeshore walk/cycleway linking the Ontario Place and Harbourfront areas.

Transport

The shift in population distribution towards the regions around the City of Toronto has had a number of implications for city planners. Perhaps the most obvious problem has resulted from increased **commuting** from these outer suburbs into the city, particularly to the CBD where most jobs are still concentrated despite moves towards decentralisation. The population of the outer regions is heavily dependent on the car, with most families having two or three vehicles to cater for their needs.

Services in the suburbs, particularly shops, schools and health care, have become road-oriented, and this further encourages **urban sprawl**.

Road transport

There are more than 1.3 million cars in Toronto, and a further 1.0 million are driven into the city every weekday from the surrounding regions (Figure 3.24). This puts considerable pressure on the city's road system and traffic congestion is increasing. An extensive expressway system was constructed in the late 1960s which had a major detrimental impact on the environment due to noise and air pollution. During the course of road construction, the city lost a large number of old buildings which in some cases were historically valuable. The road congestion is exacerbated because there have been no new motorways constructed in the central area of Toronto since 1971, when the proposed Spadina Expressway plan, which would have bisected the CBD, was rejected after widespread public opposition. There has been a major anti-road lobby in the city for over 30 years with many people arguing that money should be spent on improving public transport.

Once car drivers have arrived in the city, they face the further problem of trying to find a parking space. Although some city centre offices provide underground car parking for some employees, this is usually only available for a few top executives and the costs for other car-borne workers can be very high. Parking charges in the CBD are very high due to the high land value. Some commuters have contracted spaces with commercial car parking companies which guarantees a space for a high annual or monthly fee.

Public transport

In a deliberate attempt to combat the pressure of commuter road traffic, the Toronto planning authorities have adopted a policy of encouraging public transport at the expense of the private motorist. This policy is in sharp contrast to most cities in North America where road investment has always taken priority over public transport. In Toronto

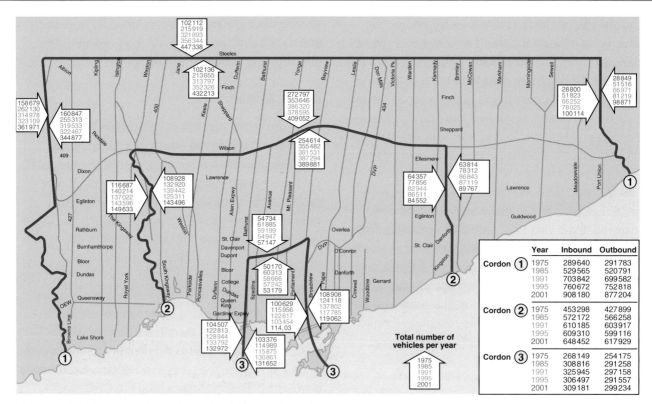

Figure 3.24 Vehicles crossing three major cordons in Toronto, 7a.m. – 7p.m. 2004

the urban transit lines form a vast network of routes linking the city's suburbs to its CBD, carrying millions of commuters in and out of the city each day (Figure 3.25). The system is administered by the **Toronto Transit Commission (TTC)** which operates buses, trolley buses, street cars and subways in a fully **integrated transport network** of over 6000 km of route. In 2003 user figures exceeded 405 million passenger trips each year. Figure 3.25 illustrates the important link that has developed between the public transport system and the move towards out-of-town office centres such as those at Etobicoke, North York and Scarborough.

The policy of the TTC is to provide public transport to within 600 m of 95% of Toronto's population. Its success has been due to the fully integrated nature of the system, for example, journeys that involve a change in type of transport do not involve additional charges and the interchanges are designed to minimise waiting and inconvenience (Figure 3.26).

Land values increase greatly when public transport is improved. For example, in the 40 years since the opening of the Yonge Street subway line, 50% of all new apartment construction in that part of Toronto was built within walking distance of the line. During the same period 90% of all new office construction took place near the major subway stations, and all of the new out-of-town centres at Scarborough, Etobicoke and North York have rapid-transit stations.

The development of the **GO train** (Figure 3.27) network, which reaches out into the developing regions outside the city, handled over 44 million passengers in 2003 year from 53 local stations. Union Station is the hub of the GO network where it links directly into the TTC subway system.

The future

A recent planning report forecast an increase of 55% in the number of daily transport trips in Toronto by 2021 and since neither the existing TTC network nor the road system could cope with this, the city is planning major transport developments on all fronts. Future plans include new subway lines and a **Light Rapid Transit (LRT)** line has recently opened in the

Figure 3.25 Transport systems in Toronto

Figure 3.26 A transit link in Toronto

Figure 3.27 GO trains in Toronto

harbour area connecting Union Station with Harbourfront, the CN Tower and the SkyDome. This LRT is intended to encourage suburban dwellers to use public transport to reach the SkyDome for matches. The road traffic chaos in the area before and after major events is a major factor in encouraging the use of public transport.

Infrastructure water and waste

As population has grown in the Toronto area, the demands on services such as water supply and waste disposal have intensified. The people of Toronto use vast quantities of water every day for domestic and industrial purposes, but within the last decade city residents have become more conservation conscious as the need for sustainable

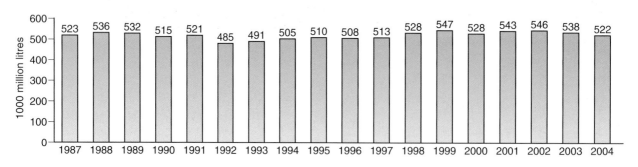

Figure 3.28 City of Toronto water consumption, 1987–2004

use of resources has been recognised through education programmes and advertisements. These have contributed to a decrease in residential per capita water use (Figure 3.28), although consumption across the city has increased due to population growth and improved economic activity.

The most successful conservation programme in the city relates to solid waste (Figure 3.29). During the late 1980s a scheme was developed to divert waste from landfill sites, as a result of which diverted waste increased by over 1000% from 21 553 tonnes in 1987 to almost 250 000 tonnes in 1998. The most innovative component of the waste diversion scheme is the city's **Blue Box System** whereby residents pre-sort rubbish before collection into separate boxes for paper, metal, plastics etc. This system has been copied by many Scottish local authorities.

Managing urban change

The major successes in managing urban change in Toronto are in part due to the changes in local government organisation. There can be no question that the Greater Toronto Area is a single urban entity,

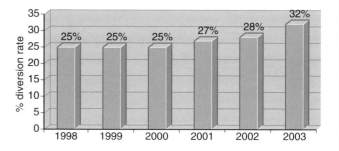

Figure 3.29 City of Toronto diversion rate of solid waste to recycling, 1998–2003

marked by an older core city surrounded by newer regional areas. As the city has expanded, there has been recognition that planning for housing, employment, transport and the environment can only be undertaken for the entire GTA area. At the same time it has been recognised that there are differences between Toronto and the GTA regions, rooted mainly in the city's complex socio-demographic mix and its mature urban form. As we have seen, Toronto still faces many social, economic and environmental challenges if it is to retain its claim to be 'the liveable metropolis'.

QUESTIONS

29 Explain the link between income polarisation and the social structure of the GTA area.
30 Describe the movement of population within and between the City of Toronto and the surrounding regions.
31 In what ways does the St Lawrence area of Toronto illustrate the city's 'liveable metropolis' claim?
32 Summarise the main elements of Toronto's open

spaces system and outline how they may be under pressure.
33 What problems face drivers in Toronto?
34 Explain how the TTC's integrated transport system helps to resolve traffic congestion in the city.
35 What successes have been achieved in managing public services such as water supply and waste disposal in Toronto?

Urbanisation in Pakistan

Understanding the current urban processes at work in an area really requires a look back at the past. This applies equally to developed and developing countries, even though the timescales of **urbanisation** may vary enormously.

The present-day boundaries of Pakistan (Figure 3.30) contain the locations of the oldest urban settlements in the Indian subcontinent (Figure 3.31) of which Harappa and Mohenjo Daro are the best known.

These civilisations, based on the agricultural potential of the Indus basin, date back to 3000BC, with urban development at its peak from 2000BC to 1700BC.

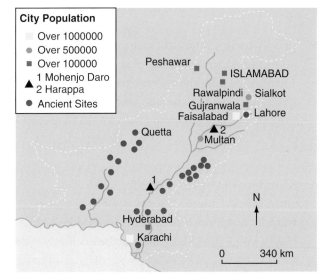

Figure 3.31 Urban centres in Pakistan

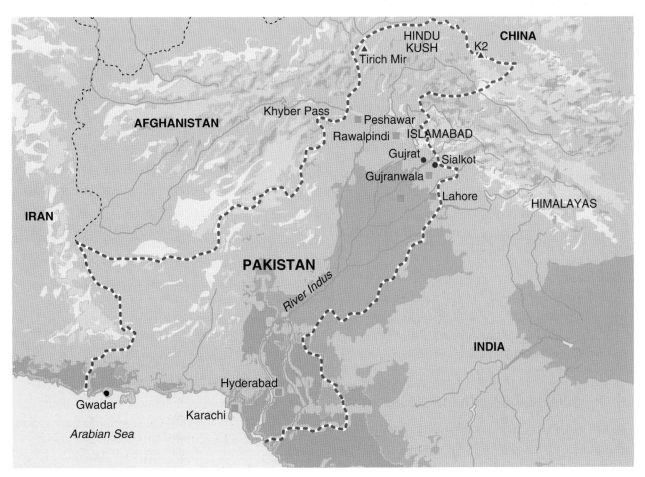

Figure 3.30 Satellite image of Pakistan

Phases of development

The main phases in the development of the modern cities of Pakistan are shown below. In each case, the examples chosen are cities that can trace their origins back to the particular phase under which they are listed.

◆ **Pre-historic (3000BC)** Harappan civilisation has been traced through archaeological findings. There was a network of craft towns and a settled agricultural system based on irrigation. Settlements like Mohenjo Daro are no longer in use but the original site of Lahore dates from this era.

◆ **Kingdoms (AD1000–AD1200)** Cities like Qureshi and Lahore grew up as trade, religious and administrative centres for early Buddhist and Hindu kingdoms. Many of these early cities were partially destroyed by subsequent rulers and then saw a rebirth in their fortunes 200 years later in the Muslim period.

◆ **Muslim Era (AD1500–AD1800)** Many of the cities of the early kingdoms dropped out of historical sight as they were attacked, devastated and abandoned. The Mughal Empire conquered the Punjab in the early part of the sixteenth century and this era saw a rebirth of the older settlements and the growth of newer towns. Quetta and Peshawar date back to this period.

◆ **Colonial (1840–1947)** Under the domination of the **British Raj** there was an expansion of settlements linked to the development of ports, railways and the military. Karachi and Hyderabad are the main cities from this era.

◆ **Post-colonial (1947–present)** Urban development was linked to the production of five-year plans, a legacy of the bureaucracy from the colonial period. The new capital of Islamabad was constructed as an exercise in town planning.

Urban population growth in Pakistan

It is vital to set the urbanisation process against the background of the overall growth of population in the country (Figure 3.32).

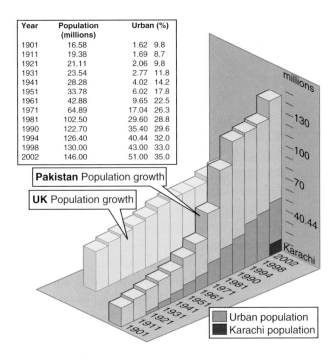

Year	Population (millions)	Urban (%)	
1901	16.58	1.62	9.8
1911	19.38	1.69	8.7
1921	21.11	2.06	9.8
1931	23.54	2.77	11.8
1941	28.28	4.02	14.2
1951	33.78	6.02	17.8
1961	42.88	9.65	22.5
1971	64.89	17.04	26.3
1981	102.50	29.60	28.8
1990	122.70	35.40	29.6
1994	126.40	40.44	32.0
1998	130.00	43.00	33.0
2002	146.00	51.00	35.0

Figure 3.32 Population growth in Pakistan, 1901–2002

The first full census of India (which at that time included present-day Pakistan) took place in 1871. Wherever possible, the figures for Pakistan are projected back to allow a clearer picture to emerge. One of the most significant trends noted from a study of these figures is that Pakistan is one of the few countries in Asia with a population growth rate as high as 3% per annum.

From Figure 3.32 it is also clear that while the urban population of Pakistan is only around 33% of the total population at present, the growth rate of this urban population is running at a higher rate than the overall growth rate for the country, thus providing statistical evidence for the process of increasing urbanisation. While the evidence of increasing urbanisation is clear, it is more important to try to understand the causes of the process, and to look at the results in terms of the effects on the people and their country.

The root causes of the urbanisation process lie predominantly in the rural areas, where the 'push' factors drive increasing numbers of people, both seasonally and permanently, in search of employment and accommodation in the cities. Many of Karachi's immigrants are in fact itinerant agricultural workers who come into the city during those times of the year when they cannot find employment on farms in the surrounding area. Eventually, and especially during prolonged periods of drought, many will settle permanently in the city, thus swelling the numbers in the **katchi–abadis (squatter settlements)**. In order to stem the flow of migrants, it is necessary for the government to address the problems faced by farmers in the rural **hinterland**, and to seek to establish alternative means of employment for them in suitable industries.

The results of increasing urbanisation are clearly evident in the cities themselves, where perpetual change and the burgeoning population complicate an already difficult management situation for the politicians and planners. Despite the problems faced by many cities in the developing world they continue to exert a powerful attraction for those seeking a way out of the poverty and misery of the rural areas, so clearly many 'pull' factors continue to operate. These include the potential for employment and the expectation of accommodation, both of which all too often fail to materialise.

Urban trends in Pakistan

The smaller urban centres in the northern regions of Peshawar and Rawalpindi (Figure 3.31) show the greatest rates of recent growth. This is due to the magnet of Islamabad, which is a new component in the urban system in Pakistan. This growth reflects the increasing economic importance of the north, and the current relative decline in the importance of Karachi and south-west Pakistan, indicative of a long-term shift in the political geography of the state.

Urbanisation in Pakistan can be summarised at two levels:

Regional 'push' and 'pull' factors operate in and around the major cities and influence rates of growth.

Local (city level) – the character of urbanisation is shaped by the incoming population. Its requirements shape urban planning policies, and these change rapidly as the character of the population alters.

As is shown in the following case study, the planners in Karachi have been forced to accept the continued existence of squatter settlements, to manage their improvement and control their development, rather than try to remove them altogether (Figure 3.33). In this experience, Karachi exhibits characteristics that are very similar to those in many other developing cities.

Figure 3.33 Karachi skyline

QUESTIONS

1. Study the information provided in Table 3.5 (Pakistan's urban growth) and identify the trends evident from the statistics.
2. From the figures provided in Table 3.6, construct a rank order table for each statistic.
3. A United Nations population projection suggests that by 2015 Pakistan's population may reach 205 million. Outline whether you think this is a realistic projection and, if so, the likely consequences of this figure.

Table 3.5 Pakistan's urban growth

Year	Total population (millions)	Urban population (millions)	Urban population (% of total)
1981	84.25	23.5	28.3
1998	130.00	43.0	33.0
2002 (est.)	146.00	51.0	35.0

Table 3.6 Urban population in selected countries in Asia, 1998

Country	Population (millions)	% growth	% urban
India	983.3	1.7	27
Pakistan	130.0	2.4	33
Bangladesh	127.6	1.8	19
Sri Lanka	18.9	1.2	23
Afghanistan	24.7	2.5	18

Source: US Government Publication: McDevitt, Thomas M., World Population Profile, 1998

CASE STUDY
Karachi

Site, situation and structure

Karachi is situated on the shores of the Arabian Sea, to the north of the mouth of the River Indus (Figure 3.34). It is a relatively young city; until the early part of the nineteenth century there was only a small fishing village at the mouth of the Lyari River, and a fortification at Manora.

In 1839, the fort was taken by the British, who then began the development of the modern city, inland from the original coastal village. They dredged and improved the natural harbour, and laid out a street plan in the area between the Lyari and Malir Rivers (Figure 3.35), thus beginning the landward expansion of the city which has continued to the present day.

Partly due to its strategic importance, and also because it was cooler than the original capital at Hyderabad, Karachi became the capital of Sind province in 1843. With the expansion of the railway system in the 1860s and continued improvements to the now bustling harbour, its commercial and

Figure 3.35 The early growth of Karachi

Figure 3.34 The location of Karachi

industrial future was assured. This was despite the economic dominance held by Bombay throughout this period, right up to **Partition** and the creation of the state of Pakistan in 1947.

When the new state was created, the capital was established in Karachi (though this was to move to Islamabad in 1971). However, the initial impetus this gave to the commercial and industrial life of the city ensured that Karachi quickly became the largest and most important city in Pakistan, with the largest port, a major international airport, main rail and road termini and a wide variety of industries.

The **Metropolitan Area** now houses 6% of Pakistan's total population, and 22% of its urban population. It generates 15% of the nation's Gross Domestic Product (GDP), and accounts for 35% of employment in large-scale manufacturing. Karachi's rapid population growth can be seen in Figure 3.36. Most of the current growth stems from immigrants from the surrounding agricultural hinterland, or from farther afield in Pakistan. Many immigrants previously arrived in the area as refugees from India,

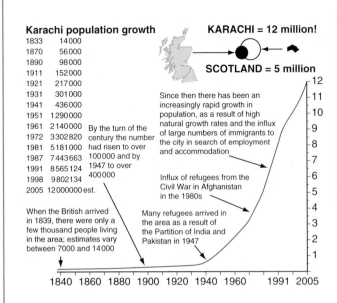

Karachi population growth

1833	14000
1870	56000
1890	98000
1911	152000
1921	217000
1931	301000
1941	436000
1951	1290000
1961	2140000
1972	3302820
1981	5181000
1987	7443663
1991	8565124
1998	9802134
2005	12000000 est.

KARACHI = 12 million!

SCOTLAND = 5 million

By the turn of the century the number had risen to over 100000 and by 1947 to over 400000

Since then there has been an increasingly rapid growth in population, as a result of high natural growth rates and the influx of large numbers of immigrants to the city in search of employment and accommodation

Influx of refugees from the Civil War in Afghanistan in the 1980s

When the British arrived in 1839, there were only a few thousand people living in the area; estimates vary between 7000 and 14000

Many refugees arrived in the area as a result of the Partition of India and Pakistan in 1947

Figure 3.36 Population growth in Karachi, 1840–2005

1900 (original core)
1947
1965
1978
1989

0   10 km

Figure 3.37 The modern growth of Karachi

as a result of the Partition of the two countries which coincided with Independence in 1947. As a result, Karachi now has a very mixed population; the Sindis, natives of the region, account for less than 10% of the population, while Pathans, Punjabis, Muhajirs and Baluchis together constitute over 90%.

By the time of the 1998 census, the population of Karachi had risen to almost 10 million (Figure 3.36), and estimates for the year 2005 are put at over 12 million. By the year 2015, the United Nations has predicted that this figure could rise to over 20 million.

The physical growth of Karachi

The present city of Karachi covers an area of about 150 000 hectares (Figure 3.37). In 1947, the total built-up area was only 18 600 hectares, and most of that was due to rapid expansion under the direction of the British colonial administration in the late nineteeenth century. The original site consisted largely of a group of islands and sandspits. On the landward side were some barren hillocks, and it was between these and along the shore of the main bay that the early fishing villages grew up.

By the time the British arrived a small walled town had developed in what is now the commercial heart of the city, stretching from Kharadar near the shore

to Mithadar on the Lyari River. From this centre expansion has taken place mostly to the north and east. Most of the older public buildings date from the late nineteenth century, during the period of British colonial rule.

Many of the housing areas on the fringe of the CBD were built to accommodate military personnel and civil servants, and were known as Barracks, Quarters or Lines, names which they retain to this

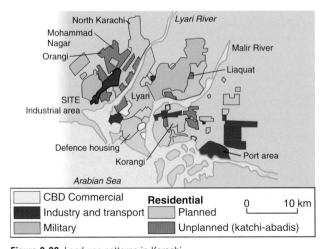

CBD Commercial	**Residential**	
Industry and transport	Planned	
Military	Unplanned (katchi-abadis)	

0   10 km

Figure 3.38 Land use patterns in Karachi

day (Figure 3.38). The Pakistani Army still provides accommodation for its officers in the city in this way.

Managing change in Karachi

The Role of Authorities

Karachi Municipal Corporation (KMC) is the prime local body responsible for municipal affairs of the city. The major functions and responsibilities of KMC include:

◆ general maintenance of urban areas, including landscape, parks and recreational grounds
◆ provision of community services such as hospitals, dispensaries, maternity homes, basic health care units, schools, colleges and basic education
◆ leasing and public property maintenance of roads, streets, bridges, markets, public buildings
◆ construction of roads and flyovers
◆ co-ordination of planning and development activities
◆ solid waste management, including sweeping, collection and transporting to landfill sites
◆ firefighting services
◆ squatter site upgrading / management and leasing
◆ flood control, maintenance of street lights, slaughter houses and graveyards
◆ sanitation and maintenance of stormwater drainage.

Karachi Development Authority (KDA) was formed in 1951 as the Karachi Improvement Trust. In 1957, it was restructured as the Karachi Development Authority. The authority is **autonomous** in nature and performs the following functions:

◆ master planning, co-ordination, environmental control, urban design and urban renewal
◆ land management and development control
◆ maintenance of **infrastructure** in the schemes managed by the development authority.

Defence Housing Authority was established in 1953 under the Ministry of Defence. The authority

provided residential plots to the retired and serving officers of the armed forces personnel. It now owns the most expensive residential localities of the city and manages an area of around 6500 hectares.

Karachi Port Trust, formed in 1887, is responsible for the maintenance and development of the Karachi port and adjoining ancillary estates. It also provides housing facilities to its employees in the immediate localities. Planning decisions and the management of change in a city like Karachi pose particular problems for municipal authorities. The city is growing at a faster rate than the politicians and planners can cope with. The Karachi Development Authority has to work in a reactive way. It can make five-year plans, but because of lack of money and the control to implement the plans they may stay on the drawing board. In Table 3.7 the problems for the authorities in Karachi are compared with a city in the UK.

Karachi: squatter settlements

One of the characteristic features of many large cities in developing countries is the existence of large areas of poor-quality **shanty housing** or squatter settlements, often on the outskirts of the city (Figure 3.39), which lack any of the basic amenities such as piped water supplies (Figure 3.40), sanitation systems or electricity. In Karachi, these are known as **katchi-abadis** (Figure 3.41). There are 536 listed katchi-abadis with an estimated population (2000) of over 7 million people, and rising.

Figure 3.39 Land to be developed in Karachi

They have come into existence partly as a result of the lack of proper urban management and planning, and partly in response to the lack of adequate affordable housing for the poor in the city. The city authority does not have the power to prevent the establishment of such areas.

A typical katchi-abadi

Lyari is an established squatter settlement, situated in the delta of the Lyari River. It has 600 000 inhabitants and is one of the largest settlements.

The area is becoming more and more densely

Table 3.7 A comparison of planning, budgeting and services in the UK and Pakistan

UK	Pakistan
Planning	
In a city in the UK the municipal authority will have a highly-qualified and well-trained planning department. The city will be divided up into defined areas and Local Plans will be drawn up at regular intervals. These provide a framework for control over land use and management of city growth and guidelines for the provision of services and **public utilities**. There is always a period of consultation before a plan is finalised.	In Karachi there are several different bodies which have authority over aspects of town planning. The KDA has the role of making up the actual plans for the city. In the past they have been given help by the United Nations Development Programme (UNDP). The implementation of the plan is divided between several autonomous city agencies with no clear links. The planning and development control system is fragmented with no effective, area-wide system for controlling developments.
Budgeting	
The local authority gains its income from direct local taxation and grants from central government. A huge bureaucracy has been set up to collect these taxes and a yearly planned budget is set to provide services for its citizens.	A large proportion of city-dwellers in Karachi earn low incomes and cannot afford to pay for services which are provided. The bulk of their income will go on food and they cannot afford rent for flats built by the authorities. They are forced to build **shanty dwellings** wherever they can and in many cases these have no access to public services. Theoretical physical planning is not linked to the ability to pay for the actual scheme. In one example, if a plan by the KDA to upgrade slums in one area over a five-year period had actually been carried out, the total cost would have exceeded the total income for the whole Sind region by five times.
Services	
The local authority will allocate funds for roads, education, water, building and maintenance of council housing, recreation, sewerage, street lighting etc. Public services like the electricity and gas companies will work alongside the council in housing areas allocated by the council planners and built by private companies. There is no unplanned development and provision of public services is uniform throughout the specific area.	Service provision is not uniform across the city. There are some areas which have access to all amenities such as water, electricity, gas, sewerage, waste collection, educational and health services. In other areas people have built their own houses with no access to any of these public utilities. **Electricity** Although nearly 85% of Karachi's population is served by electricity, most districts suffer from regular cuts in supply. **Sewerage** Access to the sewerage system varies from only 12% in the katchi abadis to over 85% in the planned housing areas. **Piped water** Again the concentration rate varies from 50% in the katchi abadis to 84% in planned housing areas.

Figure 3.40 Supplying water to squatter settlements

Figure 3.41 Shanty houses in Karachi

packed as newcomers build their homes here (Figure 3.42). There is no planning at all in the area. It is ideal for jobs as it is so close to the city centre and the port of Karachi. Due to its ideal location it is sought after by industrialists and speculators. They demolish the shanty houses and build four- or five-storey tenements with a workshop or shop on the ground floor. Attempts have been made to clear the area by demolition but these have been met with local resistance.

Most of the houses are made from cement blocks but the quality is very poor and they are badly affected during rain or floods. The labourers who live in the area have to walk to work and services are limited. After a lot of pressure the municipal authorities have put in some water standpoints, some garbage dumps and public latrines. Construction and maintenance of these is very poor, and because they cannot cope with the huge numbers involved they are extremely unhealthy.

Around the public standpoints there are pools of stagnant water and mud. Garbage is not collected regularly and lies around in piles. One of the latrines had to be demolished because the smell was more than the local residents could stand.

How do katchi-abadis develop?

Figure 3.43 shows that the katchi-abadis are distributed widely throughout the city area, although there are some concentrations in particular areas.

The Lyari area near Karachi's city centre consists of many older squatter settlements which are by now very well established, while the areas around Baldia and Orangi are more recent developments in the north-west part of the city. There are three ways in which these unauthorised settlements have developed.

1 Illegal sub-divisions

When the city authorities started to try to restrict the growth of squatter settlements, particularly in the 1960s in Karachi, alternative sites were sought by the settlers who continued to stream into the city. The system of illegal subdivision of state land on the fringe of the city became commonplace. This system

Figure 3.42 Lyari, a city centre katchi-abadi

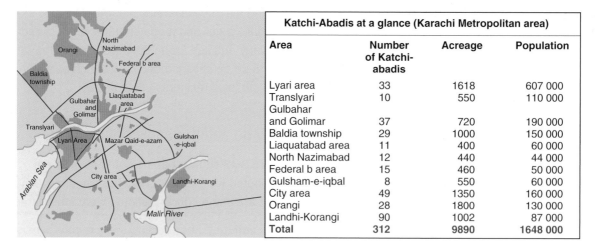

Katchi-Abadis at a glance (Karachi Metropolitan area)			
Area	Number of Katchi-abadis	Acreage	Population
Lyari area	33	1618	607 000
Translyari	10	550	110 000
Gulbahar and Golimar	37	720	190 000
Baldia township	29	1000	150 000
Liaquatabad area	11	400	60 000
North Nazimabad	12	440	44 000
Federal b area	15	460	50 000
Gulsham-e-iqbal	8	550	60 000
City area	49	1350	160 000
Orangi	28	1800	130 000
Landhi-Korangi	90	1002	87 000
Total	312	9890	1648 000

Figure 3.43 Distribution of katchi-abadis in Karachi

revolves around a middleman (or *dallal*), who acquires the occupation of state land by doing deals with local businesses, and then sub-lets plots of land to settlers at prices they are able to pay. He also arranges water supplies and protects the residents from eviction until the settlement is so big that it is safe from that threat.

2 Unauthorised invasions

Many of the earlier katchi-abadi settlements of Karachi arose as migrants to the cities arrived in groups and occupied plots of vacant land as near as possible to the city centre or their workplace. These settlements are therefore unplanned and very crowded, with narrow winding lanes, and are often the first targets of attempts to upgrade or improve squatter settlements. Due to the way in which settlers arrived, people tend to stick together in their clan structure, so there is clear ethnic as well as social grouping in these areas. The houses, which are often very similar to the rural village types, have been improved over the years, and services have been built up in a 'do-it-yourself' fashion with the help of the *dallal* and local business people.

3 Organised invasions

In some areas, due to the increase in land prices, people cannot afford to use the *dallal* system to acquire a house plot. As a result, some groups have resorted to what have become known as organised

invasions. They earmark a suitable site, occupy it in the evening and build houses on it during the night, then bribe the authorities to prevent demolition and apply for the legal rights to stay. Only a few such invasions have occurred in Karachi so far, but it seems likely that more will occur in future due to the lack of available land in convenient locations at prices people can afford.

In the early stages of growth the shanty dwellers stand a good chance of being evicted. It is often this threat to their security that encourages inhabitants to join together to resist eviction and to demand the improvement of services.

In the longer-established squatter settlements, especially those where community action has improved conditions, most of the houses are built from bricks or blocks, with proper roofs, and have a reasonable level of facilities such as water, sewerage, garbage collection, storm drainage and electricity. Shanty dwellers are more likely to be given legal rights to their homes if they have been able to provide some basic services for themselves.

The development of a typical squatter settlement in Karachi

Stage 1 (initial invasion) – A few huts appear and are often bulldozed several times before landlords or authorities are forced to recognise the existence of the new settlement. There is no piped water, no sanitation, twisting narrow streets, and no planning.

Stage 2 – Families get organised, demand services, and start to improve their dwellings. Water is delivered and sold by private contractors in large tankers. The squatter dwellers will connect their homes to an open pipe system to take sewage away from the area. As services improve there is an increase in the density of population.

Stage 3 – More improvements to dwellings take place, including the use of concrete blocks for walls, and the first **standpipes** are located in the area. The houses are provided with electricity from diesel generators.

Stage 4 – Drainage and sewerage systems are now installed, usually through self-help schemes, and many houses have water supplies and electricity (at a price). The first shops and two-storey houses are appearing, and the plots are often subdivided as rents continue to rise.

4 Why is it so difficult for city authorities to stop the growth of katchi-abadis?
5 Describe the role of the *dallal* in the creation of katchi-abadis in Karachi.
6 Describe the major differences between the planning systems in a UK city and in Karachi.
7 In what ways has the growth of population in Karachi been increased by political factors outside the control of the city authorities?

House types in a katchi-abadi

In one analysis of the squatter housing in Karachi, seven house types were distinguished. These can be seen in Figure 3.44.

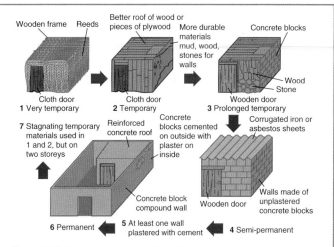

Figure 3.44 House types in a katchi-abadi

Most residents of squatter settlements finance the building of a house from what savings they have, although some are forced to borrow money from friends or moneylenders. In some cases the *thallas* (makers of bricks and blocks), on whom the community depends for building materials, will supply them on credit. This is widespread in the katchi-abadis in Karachi.

Most houses begin with one room and a toilet and this is added to over the years so that after ten years the house may have four or five rooms and it may be more than one storey high. Although masons and other tradesmen are employed by most residents, standards are usually poor and the buildings suffer from defects in structure and poor ventilation.

Sewerage is the biggest problem in katchi-abadis, and is often the most neglected service. In Karachi the systems drain into the *nullas* (natural gullies), where the effluent gathers and slowly soaks away. The health hazards from these practices are serious, particularly where the groundwater is contaminated by these soaks. Health care in katchi-abadis is almost exclusively provided in private clinics, which are usually run by unqualified staff who often use out-dated drugs to treat patients.

Most children are educated in private schools, run on a purely commercial basis and geared to what the residents can afford. They are therefore poorly built, inadequately ventilated and lit, with poor furniture and resources, and staff are almost all untrained, many of them semi-voluntary.

Services in a katchi-abadi

Water

In the early stages of development, water is supplied by bowzers (large tankers), usually from the Karachi Metropolitan Corporation via *dallals*. Payments are arranged through the *dallal*, who pockets a profit from the deal. As the settlement grows, these functions are increasingly carried out by local entrepreneurs, who use donkey-carts to deliver water from standpipes directly to customers.

Electricity

Electricity is provided to those residents who can afford it, usually from diesel generators which are installed and operated by local entrepreneurs. The service is only offered for a few homes each evening and on hot afternoons.

Transport

People also begin to make their own arrangements for transport; this includes using small Suzukis to carry passengers to and from work, and to and from local bus stops. The operators have no licence or permits, but have come to an 'understanding' with the local authority representatives (Figure 3.45).

Sewerage

In many areas, it is through the efforts of the people themselves that improvements are made, especially when it comes to the installation of sewerage systems. Initially simple tools are used to run small pipelines to the nearest *nulla* (natural gulley) or to a riverbed. However, these efforts are often wasted since the pipes are not strong enough to stand up to

Figure 3.45 Transport in Karachi

the use made of them, they are not part of an overall planned system and they quickly block up or fall into disrepair. Since there is an obvious health problem attached to this situation, it is a key area for organised action to take place. The following example illustrates one way out of these difficulties for katchi-abadi dwellers.

The Orangi Pilot Project

In the north-west part of Karachi there is a large unauthorised settlement called Orangi, or Orangitown (Figure 3.38). It contains 750 000 people, and has few basic services. In 1980 a self-help programme called the Orangi Pilot Project (OPP) began, with official support, to try to help the communities in this settlement area to develop and improve basic services, particularly sanitation.

The Low Cost Sanitation Programme of the OPP is its most successful effort. Through it the OPP has motivated the residents to manage, finance, operate and maintain an underground sewerage system. The OPP designed the system, worked out its costs, and provided tools and supervision while the residents organised themselves, collected the required funds and managed the implementation of the project. Technical research, along with the elimination of contractors, have lowered costs to a quarter of the KMC rates for similar work. As the people have funded the work they also maintain it.

Most of this maintenance is in response to a crisis, such as blockage of a drain resulting in flooding of a neighbourhood. In some cases people have, however, developed proper neighbourhood organisations that take care of maintenance and charge the residents a regular fee for it. However, the system eventually flows into the open *nullas*, the development of which is beyond the financial and organisational capacity of local people. The OPP has so far assisted the people in providing underground sanitation for about 60 000 housing units. Further extensions of all of these services will depend very much on the efforts of the *dallals*, who can often succeed in getting political support (in exchange for promised votes, for example) for improved water supplies and proper electrical connections.

Employment in the katchi-abadis

The residents of squatter settlements have chosen to live there in order to be near to the only available types of employment open to them, which is why most have come to the city in the first place. In Karachi many of the residents of squatter settlements are employed in the chemical and textile industries, and they are used as a cheap source of labour. Many have little knowledge of their legal rights in terms of conditions and wages, and many work in dangerous surroundings. Many of the women gain work in the garment industry, both in cramped factories and at home (doing piece-work). Similarly, shoe-making is carried on in many katchi-abadis, but it is not mechanised and there is little opportunity for profitable expansion.

Families may keep buffaloes and cows, selling the milk produced from them. Some of the women also make fibre mats, ropes and brooms, with raw materials supplied by a middleman who takes most of the profits. There is a thriving business in the recycling of garbage, with glass, plastics, metals and other materials being collected mostly by children, who also pass them on to middlemen for distribution to manufacturers.

QUESTIONS

8 In what ways is service provision in the katchi-abadis never likely to keep up with population growth?
9 Describe the different ways a katchi-abadi dweller might improve on an original reed hut.
10 Why are there serious health problems in a katchi-abadi?
11 Describe the ways in which services are improved in a katchi-abadi.
12 Describe what the residents have to do to get any services in their katchi-abadi.
13 What are the advantages and disadvantages of the self-help schemes in a katchi-abadi?
14 Describe how the assisted self-help scheme has improved services in Orangi.
15 Describe how the Orangi scheme operates and how it offers a cheaper and more effective method than that normally offered by the municipal authorities.

Urban differences

So far, this case study has concentrated on the problems of the katchi-abadis of Karachi, in particular the Mohammad Nagar area of Orangi township in the north-west part of the city. However, Karachi also contains a number of areas of middle-class and high-income accommodation, many of these dating back to the period of colonial or military domination in the days before Partition (Figure 3.46).

Figure 3.47 shows the street plan of part of one such area. The Defence Officers Co-operative Housing Society was set up in 1952 as an independent organisation with the aim of re-housing military personnel during the chaotic years immediately after the Partition of India and Pakistan. The Society has the legal right to completely control the organisation and planning of the urban landscape in the area under its jurisdiction. Table 3.8 highlights the contrast between this area and katchi-abadis such as Mohammad Nagar.

Figure 3.46 Upper-class housing in Karachi

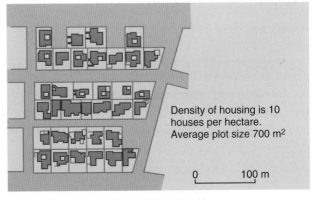

Density of housing is 10 houses per hectare. Average plot size 700 m²

0   100 m

Figure 3.47 Defence Society housing in Karachi

Table 3.8 Urban areas data

	Defence Society	Mohammad Nagar
Total area (ha)	6.5	4.4
Number of plots	67.0	230.0
Number of houses	67.0	300.0+
Density (house/ha)	10.0	62.0
Public area (ha)	2.0	1.3
Private area (ha)	4.5	3.1
Average plot area (m^2)	700.0	135.0
Average built-up area (m^2)	285.0	115.0
Average garden area (m^2)	415.0	91.0

Source: Masterplan 2000 – Karachi Development Authority

There could hardly be a more marked contrast within the boundaries of one city. There are obvious differences in the types and layout of the houses. There are also differences in the use of the streets. Newer katchi-abadis tend to have a wider planned street layout than the older katchi-abadis nearer the centre of town. In Mohammad Nagar, however, the streets themselves are little used; this is because no one owns a car, whereas the norm in the Defence Housing Society is for two cars per family, and a busy four-lane road connects this area to the city centre.

Water in Mohammad Nagar is provided via thirteen standpipes, which the inhabitants had to pay for and install themselves, after waiting five years for permission to do so from the authorities. In the Defence Housing Society water is piped into every house, and daily water consumption there amounts to 80 million litres per day.

There are no medical or educational facilities at all in the Mohammad Nagar area, and very few throughout the whole of Orangi township. The cost of travel to other areas is prohibitive, so few children go to school, and most people are forced to receive medical help from illegal practitioners.

Improving housing in Karachi

Any improvement in the housing situation in Karachi is likely to be achieved by co-operation between the municipal government and the residents themselves. The Pakistan government in the 1960s and 1970s did not have a clear housing policy and did not plan housing programmes with a set amount of finance. During this time there were some projects in response to immediate need. Slums were cleared and low-cost flats were built. But even these were beyond the reach of the poorer sections of society, and many were built too far away from available jobs. Such projects were not well planned as far as finance was concerned. The low-income groups could only afford them if they were subsidised and this in turn used up a large amount of the budget and did not provide enough return to finance future projects. There was no way this sort of planning could continue in the long term, as all it did was further drain a limited budget. Some of the projects are listed below:

1950–51 Karachi Improvement Trust was formed to deal with the different urban problems.

1958 A martial law form of government was established which decided to move the poor out of the city. The **Greater Karachi Resettlement Plan** aimed to resettle 300 000 refugees in high rise flats. Only 10 000 were resettled by 1964, after which time the plan was shelved.

1968 A master planning exercise was initiated and the study revealed that 1.5 million people belonged to low-income groups.

1974–85 The Karachi Master Plan recommended the provision of 91 000 housing plots by 1980 through 'Metroville Housing Project'.

1978–81 The development authority developed three major townships containing a total of 164 891 plots.

The inadequate and inappropriate response of government policies towards housing problems led to the creation of informal sector housing, with the following problems:

- absence of legal titles to land
- delays in regularisation of ownership
- inadequate physical and social infrastructure
- dilapidated housing conditions
- evictions
- danger from disasters, especially in the flood-prone areas or low-lying lands
- absence of representation at the policy-making level.

Karachi 2000 Project

The KDA realised it did not have the expertise to produce a development plan for land use in Karachi. Help was sought from the United Nations Development Programme and the World Bank. The UNDP provided six experts and a joint American–Czech planning firm was contracted to execute the planning work under the title of 'Karachi 2000'. They drew up three basic housing development programmes which were recommended for the lower-income groups in the katchi-abadis.

1 Improvement and Regularisation Programme (IRP)

This was the plan for the lowest-income groups, and involved the following:

- A survey of all unauthorised katchi-abadis to see if they could be improved.
- Secure land tenure. Most of the katchi-abadis were built on land belonging to the municipal authorities. If improvement could be carried out, land ownership could be transferred to the residents. This would give them security so that they could further improve their homes without fear of eviction.
- Improvement of public utilities to a basic standard.
- Development of a financial plan to achieve the first three aims.

2 Open Plot Development (OPD)

This scheme would provide:

- 72 m² plots
- secure land tenure
- public utilities

Figure 3.48 The utility wall scheme

- community facilities (schools, health centre)
- workshops
- a market place.

It provides the initial plans and street layouts and then the new residents build their own houses within the planned area.

3 Utility Wall Development (UWD)

This is a scheme that has been tried with some success in Bolivia and Brazil and is directed towards middle-income groups. In each plot of land the municipal authorities would provide a concrete plinth and a concrete core wall along with water supply and proper drainage. It would be up to the new residents to fit the toilet, bath and kitchen facilities and then add on rooms to this core wall as and when they could afford them. This was a very flexible scheme and offered the way forward for many thousands of families (Figure 3.48). Unfortunately, this plan was restricted by the lack of finance and only a few of these UWD schemes actually succeeded.

Improving transportation in Karachi

QUESTIONS

16 What are the major differences between housing and services in the middle/upper-class areas and in a katchi-abadi?

17 For Karachi, briefly describe the planning methods that have been used to try to solve the housing problems.

18 Why is transferring land ownership to a katchi-abadi resident seen as a first step to housing improvement?

19 Why is the OPD plan a good idea for katchi-abadi residents on a low income?

Karachi is a major industrial centre and port, and it therefore suffers from a great deal of traffic congestion in the city centre. In order to try to reduce this congestion, two possible solutions are being addressed – the building of by-passes and the increased use of public mass transportation.

Two by-passes are planned: one to the north will provide a faster alternative route from the ports to the Super Highway, and one to the south will cut through Korangi (a major industrial area) and join the National Highway.

Karachi's mass transportation facilities mainly consist of buses, railways and minibuses. The bus system is the most important component of the city's transport system (Figure 3.49). Buses at present are being operated by public as well as private operators. The public operators favour the standard single-deck buses while the private operators are running minibuses in addition to the standard single-deck buses. The Karachi Transport Corporation (KTC) has the responsibility for the public bus

services while the Karachi Bus Owners Association (KBOA) provides the private bus services in Karachi.

Private car ownership increases around 18% every year, as opposed to a 12% growth rate in public buses. There are 37 private cars for every 1000 people compared with 1.4 buses or minibuses for the same number of people.

The measures taken by the authority to attempt to improve this situation are:

◆ improve the existing bus system
◆ construct a new metro rail system in a ring around the CBD
◆ build an exclusive transit way in selected corridors for public transit modes
◆ improve rights of way for public transit modes in central locations of the city
◆ improve the institutional arrangements to support the mass transit programme.

The success of these schemes (which involve large amounts of public funding) will depend on the middle-income groups giving up the use of their cars and taking to the buses or trains. This is not likely to happen to any great extent in the near future.

Improving infrastructure in Karachi

Water supply and sanitation

Water is the most essential service and according to the estimates only half the total population of Karachi receives an adequate water supply. Industrial and commercial users are frequently affected by water shortages and water losses during distribution. A large number of household water pipes are undersized, blocked or broken, thereby limiting the performance of the sanitation system. The problem of using bucket latrines or direct drain discharge is serious in the squatter settlements of Karachi, where the bulk of the raw sewage is spilled into the sea through the Lyari and Malir Rivers. The Orangi Pilot Project boasts a successful sanitation project.

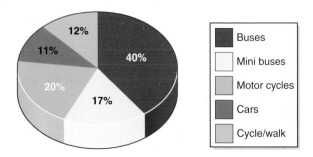

Figure 3.49 Transport modes in Karachi

- Buses — 40%
- Mini buses — 17%
- Motor cycles — 20%
- Cars — 11%
- Cycle/walk — 12%

Innovative sewerage in Karachi squatter settlements: Orangi Pilot Project

Orangi, a squatter settlement located on the outskirts of Karachi, had primitive forms of sewage disposal, poorly laid drains and no rainwater drainage. This created a dangerously unsanitary environment that led to disease, death, and damage to housing stock. Dr Akhtar Ahmed Hameed Khan, a world renowned community organiser, realised the possibility of installing low-income systems for sanitation. The Orangi Pilot Project (OPP) started using the organisational capabilities of the local leadership in each lane. The modern sanitation system was broken down into its various components:

◆ sanitary latrine inside the house
◆ underground sewerage lines with manholes and house connections in the lane
◆ secondary or collector drains
◆ main drains and treatment plants.

The emphasis of the OPP's approach was to drastically reduce the cost of construction and to persuade house-owners to accept full responsibility for implementing it. The design of manholes and shuttering was simplified so that it could be installed by the owners themselves. The 'lane manager' acted as the contractor representing about fifteen houses and the lane became the basic unit of planning. The lane managers were responsible for motivating the households, collecting the money, receiving tools from the OPP and organising the work.

In the beginning the lanes located close to the drain or the creek participated in the programme. The lanes away from the creeks had to come together to lay secondary drains to reach the creeks. This was difficult and took much longer. In the meantime sewage lines clogged up occasionally and had to be cleaned out. To overcome this problem it was decided to put a one-chamber septic tank between every connection. This mini septic tank prevented solid matter from flowing directly into the drains. The size and design of the septic tank were determined by its cost to the user and not by any engineering standard. Secondary drains were eventually introduced once more of the community participated in the programme and it became feasible to link all the lanes into it.

(Source: Akhtar Badsha, *Sustainable and Equitable Urban Environments in Asia*)

Solid waste

About 6000 tonnes of solid waste are generated in Karachi Metropolitan Corporation (KMC) every day. Only 30% of the volume is collected and disposed of each day while the remainder is either burnt or finds its way onto vacant plots, drains or roads.

Karachi is divided into ten solid waste collection districts of which seven are served by KMC while the remainder are taken care of by the airport, Defence Housing authority and the Karachi Development Authority. KMC spends about 30% of the municipal budgetary allocation on solid waste removal.

The Karachi Development Plan 2000 stresses the involvement of the private sector in collection and disposal services, and aims to promote recycling and the re-use of landfill sites. A project to dispose of garbage through a garbage train is also underway.

QUESTIONS

20 How would traffic congestion in the CBD of Karachi be reduced by the construction of by-passes?

21 Why do you think middle-class people are unwilling to stop using their cars to commute to work?

22 Describe the problems facing Karachi in terms of providing an adequate supply of water and sanitation facilities.

23 Describe some of the advantages of the sewerage system adopted in the Orangi Pilot Project.

Key terms and concepts

- autonomous (p. 148)
- Blue Box System (p. 140)
- bluff (p. 119)
- British Raj (p. 142)
- brownfield site (p. 123)
- Census Metropolitan Area (CMA) (p. 115)
- Central Business District (CBD) (p. 119)
- colonial era (p. 142)
- commuting (p. 137)
- condominium (p. 130)
- core and periphery (p. 134)
- decentralisation (p. 129)
- deindustrialisation (p. 133)
- edge city (p. 130)
- gentrification (p. 123)
- glacial till plain (p. 120)
- GO train (p. 138)
- 'greying' (p. 123)
- hinterland (p. 143)
- income polarisation (p. 134)
- industrial suburbanisation (p. 133)
- infrastructure (p. 148)
- integrated transport network (p. 138)
- katchi-abadis (p. 143)
- Light Rapid Transit (LRT) (p. 138)
- Liveable Metropolis (p. 134)
- Metropolitan Area (p. 145)
- Partition (p. 145)
- ravine (p. 120)
- reurbanisation (p. 135)
- shanty housing (p. 148)
- site (p. 119)
- slope stabilisation (p. 135)
- spit (p. 119)
- squatter settlement (p. 143)
- standpoint (pipe) (p. 151)
- suburban business district (p. 129)
- Toronto Transit Commission (TTC) (p. 138)
- urban fabric (p. 135)
- urbanisation (p. 141)
- urban renewal (p. 135)
- urban sprawl (p. 137)
- vertical development (p. 129)

SUMMARY

Having worked through this chapter, you should now know:

★ the distribution of urban concentrations and the patterns of urban growth and development in Canada and Pakistan

★ the physical and human factors involved in the location, development and structure of Toronto and Karachi and their respective land use patterns

★ the nature of urban change in these cities, particularly:

  – the causes and consequences of population growth, decline and redistribution

  – industrial, service and employment change

  – housing change, particularly in inner cities and shanty towns

  – changes in the quality of city life and the environment

  – the inequalities which may arise in large urban concentrations

★ the importance of managing urban change in Toronto and Karachi and the problems that can emerge as a result of economic, social, technological and political processes

★ the need for planning controls such as greenbelts, traffic management schemes, and improving the environment of shanty towns

★ environmental conflicts within and around the urban areas of Toronto and Karachi due to urban expansion, traffic problems, housing problems and industrial location.

Section 4

DEVELOPMENT AND HEALTH

Links to the core

This section involves identifying and explaining differences in social and economic development both between and within countries. It also looks at levels of health, health care and the incidence of disease in the developing world. It focuses particularly on the theme of population and builds on the ideas and concepts introduced as part of the Human core, such as population geography and rural geography. Since environmental issues such as climate are often significant in explaining differences in levels of development, aspects of the Physical core are integrated into the case studies.

This chapter also examines a wide range of international issues highly relevant to the world today. It is important to understand that whilst the case studies in this chapter focus on LEDCs, many of the issues examined also apply to MEDCs. For example, a recent study revealed that average life expectancy at birth can vary by as much as ten years depending on where you live in Scotland. Concerns are also being expressed in countries such as the USA and UK with regard to the potential impact of poor diet and lack of exercise on health. Some experts have predicted that unless such issues are urgently addressed, the situation could become so serious that in the near future children will be outlived by their parents.

What do we mean by development?

A general definition is that development is a process of change over time. The traditional view was that development was synonymous with economic growth – that is, countries and societies advanced by becoming richer. This view was the product of Western attitudes to growth. The emphasis for development was, therefore, on rapid economic growth in poorer nations so that the development, or economic gap, between them and richer nations could be closed.

The modern view states that development should not be measured solely in economic terms. Account should be taken of improvements to people's lives in other ways: in social terms by improving standards of living, diet, access to education and health care; in political areas by moving towards greater social justice, freedom of speech etc. All of these allow people to achieve their full potential by giving them greater choices in their lives. Allied to this is the idea of

sustainable growth – that resources are used responsibly, not exploited until they are exhausted – so that the needs of future generations become part of the development equation. In this view development is a complex process involving economic, social, political, environmental and cultural change.

Indicators of development

In the traditional view, indicators of development were based on economic indicators only, for example, **Gross National Product (GNP)** per capita (the total value of economic output divided by the population). This includes figures for food production, value of goods produced, provision of services, profits from overseas investments and money earned from foreign business. Using such an indicator, Figure 4.1 shows the distribution of high, middle and low income countries in the world. As would be expected, high income countries are found in North America, Western Europe, Japan, Australia

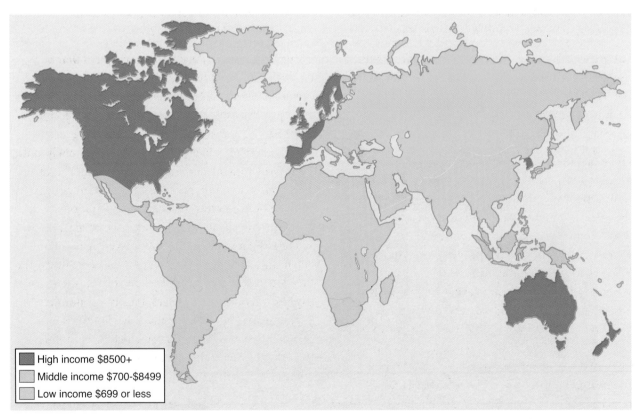

High income $8500+
Middle income $700-$8499
Low income $699 or less

Figure 4.1 Global distribution of GNP per capita

Figure 4.2 A rural landscape in the developing world

and New Zealand whilst the poorest countries are concentrated in Africa and south-east Asia.

The major advantage of using an indicator of this type is that it is relatively easy to calculate. The underlying assumption here is that if a country is wealthy the quality of life for all of its citizens will automatically be good.

There are, however, a number of disadvantages associated with the use of economic indicators. As the figures involved are averages, they disguise inequalities between the richest and poorest in society, and also between different regions within the country. The figures exaggerate the **Development Gap** (the difference in development between countries) as they do not take into account **relative purchasing power**. As the costs of goods and services tend to vary from country to country, a set amount of money may buy more goods and services in one country as compared with another where costs are higher. Economic indicators also ignore **subsistence activities** such as farming and the informal economy, despite the fact that both of these are the basis of

economic activity in rural economies in the developing world (Figure 4.2).

Using economic indicators to compare different countries also presupposes that currency values remain relatively stable in relation to one another, for example, that £100 will always be exchanged for $150 or 130 euros. However, this is not the case. Currency values change on a daily basis, usually only by a small amount but occasionally by significant amounts, so that comparisons based on monetary values alone are often inaccurate. In addition, conditions peculiar to a country, such as long cold winters or large distances to be travelled, are not taken into account. These can influence inputs to the GNP in the form of high fuel costs or more expensive goods and services. Finally, economic indicators imply that the only way to improve people's lives is to go for Western-style economic growth, ignoring other factors that might influence a country's growth.

The modern view believes that there is a need to take into account a much wider range of data, some

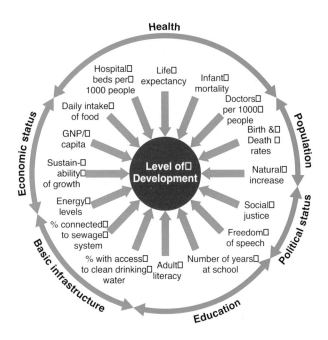

Figure 4.3 Indicators of development

of which is extremely difficult to quantify, in order to give a more rounded view of the quality of life within a country. There is a great variety of such indicators, as shown in Figure 4.3. These include statistics on health such as life expectancy at birth, infant mortality, numbers of doctors and hospital beds per 1000 of the population, and the daily intake of food. Education levels are shown by literacy levels and number of years in full-time education. Both of these groups are valuable in showing the ability of a country's government to supply basic services to its people. Population characteristics – birth and death rates and natural increase – indicate the rate at which the population is growing. A high growth rate could cause severe problems for a country struggling to develop. Economic status is shown by indicators of industrial and agricultural output, the percentage employed in the primary, secondary, tertiary and quaternary sectors and the rate of energy use. The ability of the government to supply basic infrastructure services is shown by access to clean drinking water and the number of people connected to sewerage systems. The levels of political freedom can be shown by statistics on freedom of speech and access to justice within the country.

Statistics on education standards and health care provision are relatively easy to obtain. It is much more difficult to measure such factors as freedom of

speech, or to assess whether growth is sustainable. Three indicators have been regarded as especially significant:

◆ **Life expectancy at birth** – this provides a good reflection of the quality of health care and levels of hygiene and sanitation available, and it reacts quickly to improvements in the provision of such services.

◆ **Access to safe water** – this indicates the quality of the provision of basic services within an area and is especially useful when studying urban/rural contrasts.

◆ **Adult literacy** – this is a good indicator of the levels of skill and expertise within a country which would allow it to move forward in its development.

What is important, however, is that a *single* indicator should never be used to determine development, as a country might score well in one but badly in another. It is always best, therefore, to use composite indicators.

Composite indicators of development

There are now a number of **composite indicators of development**. One of the first to be used (in 1977) was **PQLI – Physical Quality of Life Index**. It combines figures for **life expectancy**, **literacy rates** and **infant mortality rates** so that it concentrates on social rather than economic development and was, therefore, useful as a counterbalance to the emphasis on economic development prevalent at the time. To calculate it, figures for each of the three factors for a country are entered on a scale from 0 (worst figure) to 100 (best figure). The three figures are then averaged to give the country's PQLI. A figure of 77 and above is considered satisfactory; below that the quality of life is regarded as poor. As can be seen from Figure 4.4, the PQLI distributions broadly correlate with GNP distributions, although it does throw up some interesting anomalies. For example, several middle income countries such as Argentina and Chile score better than would be expected, while others such as Saudi Arabia score far worse than many of the countries of Africa and south-east Asia. These reflect the different emphasis and spending priorities which these countries give to social policies.

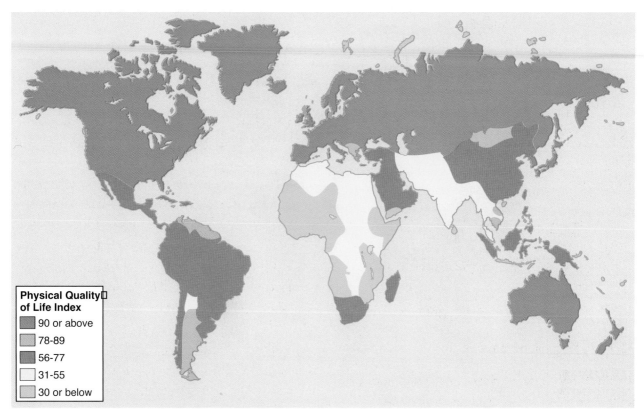

Figure 4.4 Global distribution of the Physical Quality of Life Index

A more recent composite indicator is the **Human Development Index (HDI)** which was first used in 1990 by UNPD (United Nations Programme for Development). This attempts to combine a range of economic and social indicators.

The Human Development Index uses **adjusted income per capita** (that is, the purchasing power of a given sum within the country), **educational attainment** (a combination of adult literacy rates and average number of years of schooling) and **life expectancy at birth**. The best level is 1 and the poorest is 0. A country's score for each indicator is ranked between these, and the figures added and averaged. A major advantage of this indicator is that it allows direct comparisons to be made between countries. It can also be adjusted to show figures for male and female incomes. This gives interesting results as some otherwise high ranking countries, such as Japan and the UK, slip down the rankings reflecting the disparity between average male and female earnings in these countries. Figure 4.5 shows the distribution of HDI values and indicates a broad correlation with GNP but also highlights similar anomalies to those of the PQLI. There are

limitations as only three sets of data are used. The results still mask internal differences and as these are relative figures, if all countries improve at the same rate the poorer countries will never climb up the rankings and will get no credit for their achievements.

Why measure development?

It is only by measuring development that it is possible to see what progress is being made towards improving the quality of people's lives. It helps to focus attention on areas where there is a need for more encouragement, and areas where aid from donor countries could be targeted. Comparisons can be made on a global scale, allowing analysis as to why some areas have progressed while others have not. The use of development indices has led to the classification of countries according to their perceived levels of development, for example, the **North–South Divide**, first described in the Brandt Report (Figure 4.6). Another classification used is that of **First World** (the industrialised Western democracies), **Second World** (the centrally planned

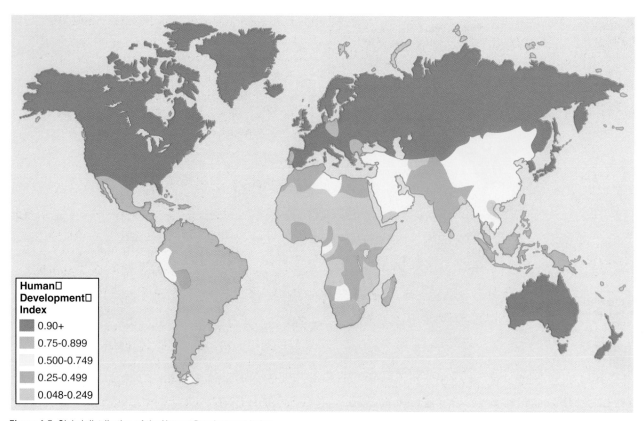

Figure 4.5 Global distribution of the Human Development Index

Figure 4.6 The North–South Divide

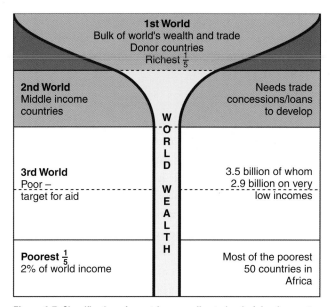

Figure 4.7 Classification of countries according to level of development

changing world circumstances, for example, the collapse of the centrally planned economies of the Soviet Union and Eastern Europe, the rise of the wealthy oil states in the Middle East and the emergence of the **NICs (newly industrialised countries)** especially the 'Tiger economies' of south-east Asia (see Figure 4.8).

Contrasts in development

There are many reasons for contrasts in development. Some countries such as Saudi Arabia or Brunei have been able to rely on income from their considerable oil and gas reserves. Other countries, for example, Ethiopia, Somalia and Bangladesh, have few or no natural resources and have experienced severe climatic problems such as drought or devastating flooding. In other areas such as Sudan and Sierra Leone, civil war and other forms of political instability affect their ability to improve their level of development. Table 4.1 shows countries with differing levels of development.

economies of Communist states), and **Third World** (poor, underdeveloped areas mainly in Asia, Africa and South America), as shown in Figure 4.7. These classifications have had to be modified in the light of

QUESTIONS

1 Explain the differences between the traditional and modern views of development. How does this affect the choice of indicators used in each case?
2 What are the advantages and disadvantages of relying on economic indicators of development when analysing a country's level of development?
3 Why is it important to refer to a range of indicators when investigating levels of development?
4 For the PQLI and the HDI or any other composite measures of development you have studied, state

three indicators which might be used in their calculation, commenting on their usefulness.
5 Explain why it is important to measure levels of development despite the problems associated with the collection of such data.
6 Describe in detail the differences in levels of development shown in Table 4.1.
7 Referring again to Table 4.1, would you categorise Saudi Arabia as 'developed' or 'developing'? Give reasons for your answer.

Table 4.1 Contrasts in development

Country	GNP per capita $	Life expectancy	% adult literate	% infant mortality
Singapore	26 730	76	91	4.0
Vietnam	240	67	90	38
Saudi Arabia	7074	70	16	29
Bangladesh	240	58	33	77

There can also be contrasts in development within an individual country. Statistics given for a country are average figures. As such they disguise differences in levels of development within a country, so care must be taken in using such statistics. There may be a few very wealthy families, with the majority of the population living at subsistence levels. There are also regional differences, with some areas more well off than others. Rural and urban contrasts are common in developing countries. Remote, isolated and scattered rural populations are less likely to have access to services such as clean water and sanitation than those in crowded urban areas. Also, rural populations are less likely to have access to education and health services. In times of crisis rural populations are less likely to receive help from international aid agencies than those in the more accessible urban populations.

The following are some of the most common factors that can influence levels of development both within and between countries:

- **Geographical isolation** – In general it is easier to provide services to the population of a small country such as Singapore than to that of a large country such as China. Within a country, some regions may be disadvantaged by their relative isolation (for example, Amazonas in Brazil).

- **Natural resources** – Some countries have vast reserves of minerals (for example, gold in South Africa), whilst others such as Ethiopia have few natural resources.

- **Climate** – Generally, areas with less extreme climates are more attractive to settlement and therefore development (for example, south-east Brazil in comparison with the Amazon rainforest, or the frequent incidence of drought in the countries of the Sahel which limits agricultural productivity, and which in turn then contributes to the relatively low levels of health and development of the countries concerned).

- **Trading links** – Some countries have developed more rapidly due to their favourable locations in terms of trade (for example, Singapore and Taiwan) and as result have attracted inward investment from overseas multinationals looking for cheap labour and access to the expanding Asian markets.

- **History** – Some LEDCs have developed more rapidly as a result of their historic links with former colonial powers (for example, India with the UK).

- **Political corruption or bias** – Certain regions of a country may lose out on investment due to corruption within the government (for example, President Mugabe of Zimbabwe has witheld assistance from areas of the country known not to support his political party). National funds may be 'diverted' into the pockets of those in power instead of being used to improve standards of living across the entire country. Investment may be focused on the capital cities to the detriment of rural areas because the ruling elite live in the cities.

- **Government investment** – Aid and assistance may be targeted towards certain favoured areas of a country, thus enabling those areas to develop more rapidly than others.

- **Unequal distribution of resources** – The bulk of a country's wealth may be concentrated in the hands of a minority of the population (for example, oil wealth in Saudi Arabia).

- **War or conflict** – Conflicts, whether civil or international, are expensive and may soak up resources that may otherwise have been ploughed into services such as health care or education (for example, the 18 years of civil war in Mozambique in the 1980s–90s).

- **Rural–urban contrasts** – Contrasts between rural and urban areas are common in most developing countries. Remote, isolated and scattered rural communities are less likely to have access to services such as clean water and sanitation than the populations living in crowded urban areas. Education and health services are also problematic in such areas. In times of crisis, such as the Pakistan earthquake of 2005, inaccessible rural populations are less likely to receive help from international aid agencies.

NICs

An **NIC** is a **newly industrialised country** (see Figure 4.8). The term was coined in recognition of

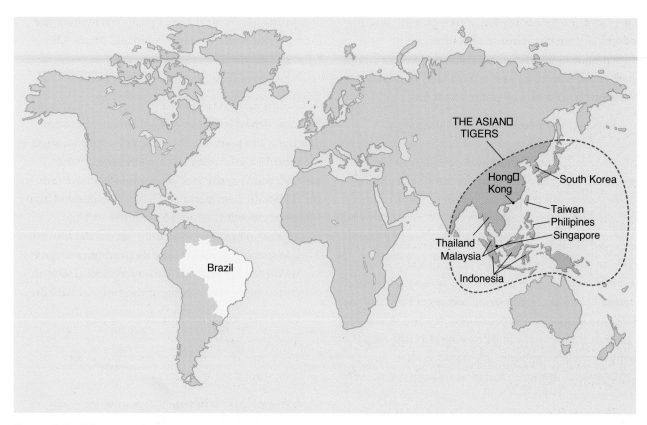

Figure 4.8 The NICs

the significant changes in the level of industrialisation in a number of developing countries. An NIC is defined as a country in which the level of **Gross Domestic Product** (the value of all goods and services produced in the country) *from industry* exceeds one-third of the *total* GDP. These changes are often due to a rapid expansion of investment by **transnational corporations** (large international businesses with headquarters in one country and a manufacturing presence in many others, for example, Shell, IBM). Included in this category are such countries as Brazil and the Asian 'Tiger economies' – Taiwan, Hong Kong, South Korea and Singapore. The last four have experienced rapid and sustained growth year on year for the last 25 years. A second wave of growth has recently occurred in south-east Asia in the economies of Indonesia, Malaysia, Thailand and the Philippines, which are attempting to emulate the original 'Tigers'.

Why have the Asian Tigers been so successful?

Some of the Asian Tigers have considerable natural resources, for example, tin and rubber in Malaysia and coal, oil, natural gas and forest products in

Indonesia. In the past these were exploited as primary exports but more recently have provided a base for manufacturing industries as the country developed. Other countries, such as Singapore, had no natural resources (apart from their harbour). However, their success was not primarily based on natural resources.

All of the Tiger countries had certain assets in common. These included a large, flexible, hard working, low-cost labour supply. Based on this it was possible to develop low skilled, labour intensive industries such as textiles, clothing, toys, plastics, leather goods and footwear.

There was also strong government intervention to encourage and help develop export-oriented manufacturing. Government policy was mainly geared towards attracting foreign investment in order to introduce high-tech industries. The policy was very successful in attracting transnational corporations which appreciated the low labour costs, reduced costs of production and created growing local markets, as the countries themselves became more prosperous. They were especially attracted by the proximity of the growing Chinese market, a lack of trade union activity and strikes, inducements from the government such as reliable power supplies, an efficient infrastructure and relaxed laws in such areas as pollution control and employment law.

In addition, local people were encouraged by government policies to save a relatively large part of their earnings so that there was considerable capital generated internally between the 1960s and the 1990s. As a result they could become major investors in their own economies to help move manufacturing forward into sophisticated, high-value new products, mainly in electronics, telecommunications and aerospace.

Consequently these countries have increased domestic production, reduced poverty and developed a high skill base and a higher living standard than many of their neighbours. The rapid growth did, however, lead to the exploitation of labour, especially of women and children, and the growth of illegal immigration from their less developed neighbours looking for work and a chance to improve their quality of life. These groups usually got the dirty,

dangerous jobs with no legal protection which the local people were no longer prepared to do. The rapid expansion also caused considerable environmental damage through deforestation, land degradation and air and water pollution.

A heavy dependence on foreign investment means that these economies are highly vulnerable to shifts in the global economy, and as a consequence they suffered badly in the world recession of the 1980s and 1990s. In South Korea, Thailand, Indonesia and Malaysia, currency values plummeted and manufacturing output was badly hit during this time. The position was made worse as overborrowing on the world money markets led to large debts, and this, combined with the collapse of the internal banking system, led to a meltdown in the value of the currency in the international exchanges. A further problem is that corruption is common at all levels within these economies and **cronieism** (where contracts are awarded to relatives and friends regardless of their competence) is a widespread practice.

CASE STUDY
Singapore

The foundation of the modern state of Singapore dates from 1819 when the island was chosen by Sir Stamford Raffles to become the centre for the British East India Company due to its natural harbour and strategic position on the trade routes between India and China. Trade was further boosted by the opening of the Suez Canal and the development of steamships (Figure 4.9).

Singapore became the centre for the export of rubber and tin from Malaysia. It grew steadily as a trading centre for the region. After independence in 1959, the new government began a crash programme to develop its manufacturing base which

proved to be a remarkable success story. In the 1960s it concentrated mainly on job creation, developing labour intensive industries such as textiles and shipbuilding (Figure 4.10).

By the 1970s, Singapore had achieved full employment and began to shift the economic emphasis to the creation of a highly skilled work-force producing high-value goods and to the encouragement of foreign investment.

Industries included shipbuilding and repair, financial services, tourism, high-tech industries (electronics, aerospace components and telecommunications equipment), textiles, chemicals, plastics, oil and petrochemicals.

The early rapid economic growth of 8% per annum is now slowing as Singapore is facing competition from neighbouring states. Its citizens have a high standard of living and per capita incomes above those of Germany (Figures 4.11 and 4.12).

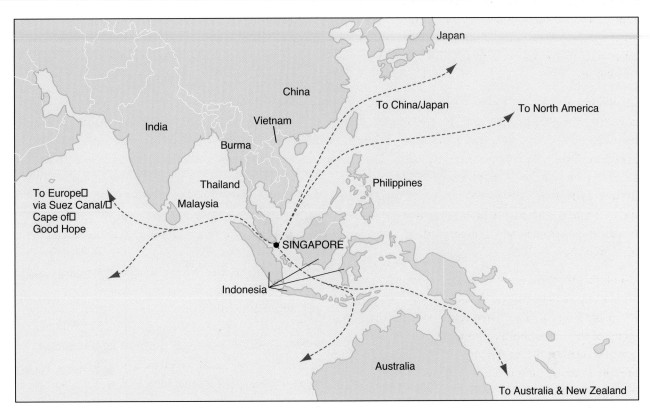

Figure 4.9 The location of Singapore in south-east Asia

Figure 4.11 Singapore skyline

Figure 4.13 Singapore harbour

All this has been made possible in a tiny country with few natural resources by a well educated, hard-working population, low population growth, excellent sea and air links and above all a stable political situation with a highly interventionist government which fosters a strong sense of national identity. The government gives strong direction towards development by creating free, open markets and ensuring that there is little corruption or crime and few strikes. This has created a favourable climate for foreign investment which is welcomed and made as free from regulation as possible (Figure 4.13).

The government's targets for the future are that Singapore becomes the new centre for international trade and a global centre for financial services (Figure 4.14), high-tech industries and business services, ranging from research and development through to manufacturing and distribution.

The country has remained stable, although the

Figure 4.10 Singapore in the 1960s

Figure 4.12 Singapore in the 1990s

Figure 4.14 Singapore skyline: central financial district

government is under pressure to relax its grip on society.

People remain unsure of the consequences of this. The economy is also vulnerable to the breakdown of government in neighbouring states such as Indonesia, which could disrupt trade. The economy is now having to face the consequences of success in the form of the high cost of land on a crowded island and the increasing costs of a highly skilled labour force. A very strong currency is making exports more expensive, and the economy has become heavily reliant on foreign investment.

The government has tried to counteract some of these problems by encouraging local companies to locate new industrial development in lower-cost neighbouring states like Malaysia, Indonesia, China, India and Burma, by building industrial estates there. Its broader, more established industrial base and proven ability to manage change has, to date, enabled it to weather the financial crisis in south-east Asia better than many of its neighbours.

Less economically developed countries

The position of the NICs is in marked contrast to the situation in many countries of Asia and Africa. There, economies have stagnated, grown slowly or even declined.

In parts of Africa growth rates have plummeted, food production has decreased, incomes per head have fallen and, overall, industrial performance has declined. It is in fact the only region of the world where per capita output declined consistently throughout the 1980s and the quality of life of its citizens deteriorated. This is highlighted by the fact that the bottom 9 countries out of 175 in terms of HDI are all in Africa.

For both the **World Bank** and the **International Monetary Fund (IMF)**, efforts to try to deal with this crisis became the focus of attention. Their solution was to insist on **Structural Adjustment Programmes** which aimed to aid economic progress. The policies tended to switch money from social action to agricultural and industrial development in an attempt to stimulate economic growth. There were also attempts at developing **import substitution** policies where local industries were encouraged to go into production in an attempt to cut down on expensive foreign imports.

These policies are now being questioned as they have not generated recovery and have inflicted considerable hardship on local populations as medical programmes and educational developments have suffered catastrophic cutbacks.

Causes of the problems in Africa

- Africa is an arid continent where, over the last 30 years, annual rainfall has been well below average – in a continent where most people are heavily dependent on agriculture. As a result deforestation, overgrazing, soil erosion and desertification are widespread. *Climate*

- Africa in general has very high population growth rates, in some countries in excess of 3% per annum. Family planning programmes have had little impact in these areas due to traditional attitudes which value large families. Such population increases tend to wipe out any progress in a country's development. *Pop growth*

- Throughout Africa there are high levels of disease, for example, Malaria, intestinal parasites, HIV and AIDS. The scale of the problem has left health systems unable to cope and led to a serious loss of members of the economically active population. There are low rates of education and many of the *Disease* small number of Africans who are educated and skilled leave for higher-paid employment elsewhere.

◆ Civil wars have created large numbers of refugees throughout the continent and led to inflated spending on military budgets at the expense of social spending. These conflicts are often a legacy of the area's colonial past when country boundaries were drawn with no consideration for tribal groupings. Such policies created political instability both within and between countries.

War

◆ In the past, exploitation of Africa's resources by colonial powers for primary exports meant that there was little development of manufacturing industry, and fertile land was used for cash crops rather than food crops. This pattern has proved difficult to change.

Exploit/resource

◆ After independence many African countries became involved in misguided, large-scale development projects which they had neither the resources nor the expertise to maintain. More recently there has been an unwillingness to be open to outside investment because of memories of past exploitation, despite their inability to

Exploit/history

generate capital for development by themselves. All these problems have been compounded by unstable governments, poor leadership, high levels of corruption and a widespread disregard for human rights.

◆ Much of Africa suffers from a poor **infrastructure** with badly maintained road and rail links. There is a lack of natural resources throughout the continent and therefore a need to import commodities such as oil. Economies were particularly badly hit by the dramatic rise in oil prices in the 1980s, a situation made worse by income from their exports falling drastically at the same time.

Infrastructure

◆ Due to heavy international borrowing during the 1960s and 1970s, African countries are weighed down by external debt, and the repayment of these debts has been taking more and more of their foreign earnings, leaving little to enable them to pursue internal development. Recently, the world's most economically powerful nations, the **G8 countries**, have begun the process of cancelling the debt of the world's poorest countries.

Debt

CASE STUDY
Kenya

In the north of Kenya there are semi-arid plains and desert which are prone to drought. In the south, savanna vegetation dominates, with many game parks, like Masai Mara. In between are the Central Highlands which are the most fertile area containing 85% of the population; population densities here are therefore very high. The area also includes the Rift Valley, a dramatic geological feature 100 km wide and 3 km deep (Figure 4.15).

Only one-fifth of the country is suitable for agriculture, most of which consists of subsistence

farms, the main crops being maize and cassava (Figure 4.16). There is limited land available for the expansion of crop area but there is potential to increase yields. In the semi-arid areas, livestock farming predominates. A succession of droughts have resulted in overgrazing, land degradation and soil erosion. There is a need to increase the area of irrigated land but this has not been possible because of lack of finance. Often the most fertile areas are used for cash crops such as tea (which provides 28% of Kenya's foreign exchange and one million jobs), coffee (the third most important foreign currency earner), fruit, flowers and vegetables (Figure 4.17). All this puts pressure on food supplies so that Kenya is now a net importer of food. The situation is not helped by Kenya's high birth rates. Tourism is increasingly important to the economy due to the wealth of wildlife and beach holidays on the Indian Ocean coast.

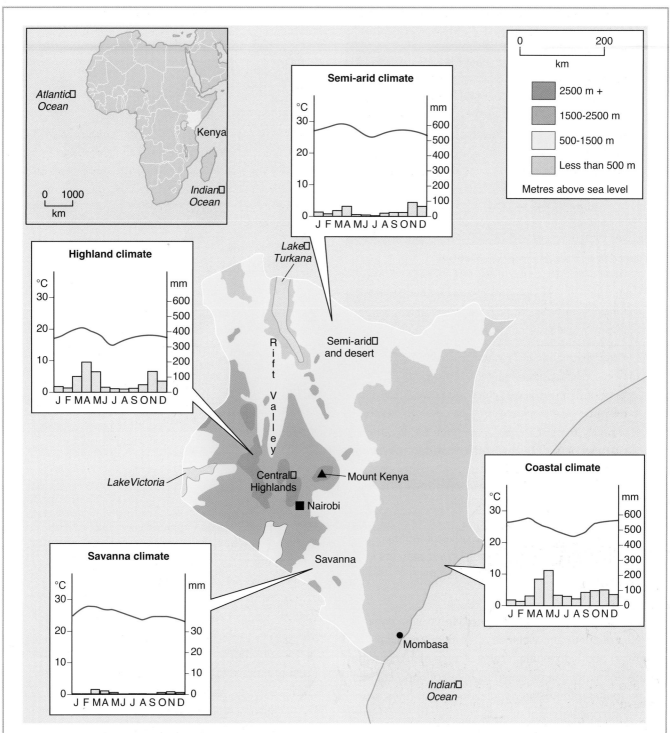

Figure 4.15 Location, relief and climate of Kenya

Despite its problems, Kenya is East Africa's most industrialised country, although the industry is based mainly in Mombasa and Nairobi. These industries supply both the domestic and export markets, and are based mainly on food processing with some chemical and engineering plants. Growth is hampered by a lack of fuel – Kenya has little coal, oil or gas and relies heavily on imports. Kenya also has to import manufactured goods and has been hit by rising prices for fuels and the falling value of its primary exports. The country suffers from frequent power cuts (as developments in hydro-electric power

Figure 4.16 Rural Kenya: a small subsistence farm

Figure 4.17 Workers in a women's co-operative

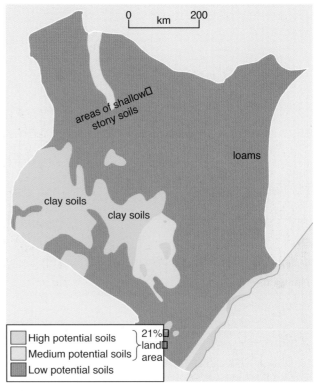

Figure 4.18 Land productivity in Kenya

are affected by droughts). The government is planning to develop geothermal power stations in the Rift Valley area and is also engaged in oil exploration.

Other problems include deteriorating roads, rising rates of crime and violence, corruption, political unrest and high inflation rates. These all badly affect the economy, put off potential investors and cut into the tourist trade. High population growth rates are making development difficult as they put pressure on education and health services, food supplies and land. Kenya needs to improve its infrastructure, especially roads, airports and water supplies (there are particular pressures on these from tourist developments as well as from the growing urban demand). Disease rates are high, especially for HIV and AIDS which reduce the capacity of the workforce in both the agricultural and industrial sectors.

Figure 4.19 Population density and communications in Kenya

Figure 4.20 Masai herders who have been forced to turn to farming

Figure 4.21 Proposed large-scale irrigation projects in Kenya

Kenya's economy is vulnerable to adverse weather conditions and fluctuations in world markets, but its government has declared its intention to become an industrial nation by 2020. In order to do so Kenya will have to create conditions attractive to foreign investment. In recent years Kenya's situation has

deteriorated markedly. The IMF suspended loans to the country in 1997 because of government corruption, although the suspension was lifted in 2000. In the meantime Kenya suffered from power shortages, inflation, water rationing and a steady fall in the value of the currency. Many farmers were unable to continue due to drought and have moved to the cities. Cash crops have also been hit by mismanagement and corruption.

The tourist industry, once a major earner of foreign exchange, is in decline because of reports of an increasingly violent and dangerous society. Despite increased aid and other international help, Kenya's economy is in such decline that it will take considerable time and effort to return it to the situation at the time of independence in 1978.

There are considerable differences in levels of development within Kenya. These differences are due to variations in relief, climate, land use potential (the capacity of the land to sustain agricultural developments) and accessibility (Figures 4.15, 4.18 and 4.19). There are also contrasts between urban and rural areas and within urban areas.

In the north and east of the country, which consists of 80% of Kenya's area, land is mainly below 1500 m and the climate is arid or semi-arid with low, erratic, unevenly distributed and unreliable rainfall. There are many areas of shallow, stony soils (Figure 4.18). All this combines to give a landscape of dry bushland, desert and scrub vegetation with a low

Figure 4.22 A coffee plantation

177

agricultural potential where, in the driest areas, harvests fail on average three years out of five.

Accessibility in the area is poor with few metalled roads and no railways (Figure 4.19). Population densities are low, with approximately 6 million people living in the area. They are mostly nomadic pastoralists (the Masai) relying on their herds of cattle and goats. Despite the poor quality of the land in the area it makes a valuable contribution to Kenya's agricultural output with 80% of Kenya's cattle found there. Population growth is high at 2.5% per annum, so there is considerable pressure on the land from overgrazing and deforestation. The area has suffered badly through a series of droughts in recent years, forcing many of the pastoralists to move to the overcrowded urban areas or to subsistence farming (Figure 4.20). The area has long been neglected by successive Kenyan governments. Despite pressure on scarce water supplies, little has been done to increase the area of irrigated land, with relatively few government sponsored schemes (Figure 4.21). There has been little or no attempt to research the agricultural problems of the region. The area is poorly served by health and family planning clinics which are scattered and difficult for many people to access.

In the south and west of Kenya, the land is mainly coastal plain or upland. Rainfall is much more reliable and soils are a mixture of clay and alluvium with small areas of fertile volcanic deposits (Figure 4.18). The vegetation is forest and savanna and in

Figure 4.24 Urban Kenya: good-quality housing

terms of agricultural potential the land is mainly in the medium to high category. Over 28 million people live in the area, so population densities are high and there is considerable pressure on land. The area produces the bulk of Kenya's crops for export such as coffee, tea and fruit (Figure 4.22).

Figure 4.23 Central Nairobi

Figure 4.25 Urban Kenya: slum housing

The area has been the focus of development with large-scale irrigation schemes planned (Figure 4.21). In addition, health and family planning clinics are more freely available here.

Most Kenyans live in the countryside (Figure 4.19); only 30% live in urban areas. There are considerable contrasts between conditions in the countryside and urban areas. Almost 70% of urban dwellers have access to clean water (which has to be paid for) compared with 50% in the rural areas. In addition, in most rural areas people have to travel considerable distances in order to collect water. In the urban areas there is access to education (which, apart from primary education, has to be paid for), doctors and other health facilities.

Within urban areas there are also contrasts. The centres of Nairobi (Figure 4.23) and Mombasa, for example, have tall modern buildings similar to those found in most of the world's cities. There are also areas of good-quality housing (Figure 4.24). Surrounding these cities, however, are urban slums which are growing rapidly as people migrate to the cities when they are unable to make a living in the countryside (Figure 4.25). In these slum areas water supply is inadequate and many people are unable to afford food or education for their children. Sewerage systems are almost non-existent and housing is of very poor quality. Disease spreads rapidly in these conditions. Many people have moved into the cities in the hope of finding employment. Unemployment rates, however, are over 50% and people find it difficult to

QUESTIONS

8 Explain the failure of the African countries to improve their levels of development.

9 Why do indicators of development not always produce an accurate picture of life within a country? Use examples from both MEDCs *and* LEDCs to support your answer.

10 Study the information on Brazil in Figure 4.26 on the following page. Using the information on Brazil, or any other country you have studied, explain why levels of development vary throughout the country.

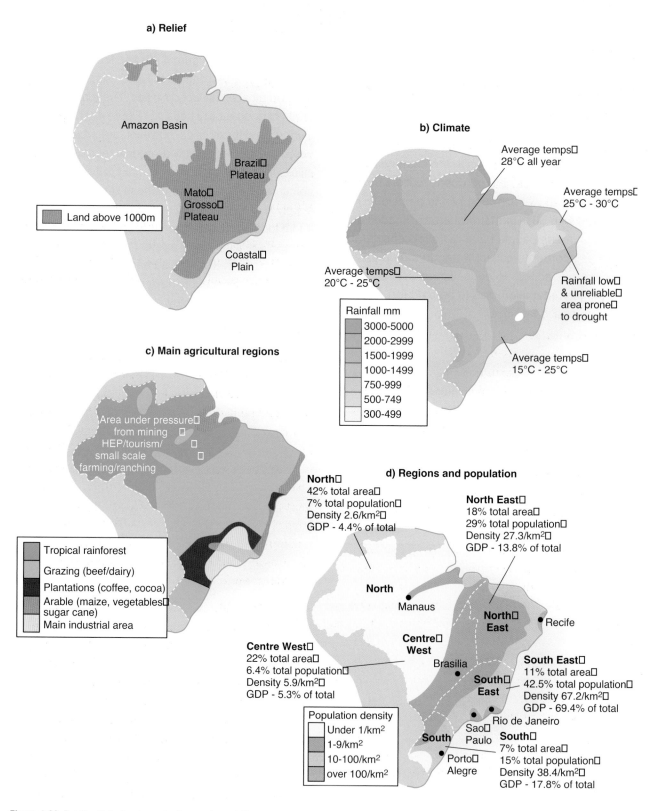

a) Relief

Amazon Basin

Brazil☐ Plateau

Mato☐ Grosso☐ Plateau

Coastal☐ Plain

Land above 1000m

b) Climate

Average temps☐ 28°C all year

Average temps☐ 25°C - 30°C

Average temps☐ 20°C - 25°C

Rainfall low☐ & unreliable☐ area prone☐ to drought

Average temps☐ 15°C - 25°C

Rainfall mm

3000-5000
2000-2999
1500-1999
1000-1499
750-999
500-749
300-499

c) Main agricultural regions

Area under pressure☐ from mining ☐ HEP/tourism/ ☐ small scale ☐ farming/ranching

Tropical rainforest
Grazing (beef/dairy)
Plantations (coffee, cocoa)
Arable (maize, vegetables☐ sugar cane)
Main industrial area

d) Regions and population

North☐
42% total area☐
7% total population☐
Density 2.6/km² ☐
GDP - 4.4% of total

North East☐
18% total area☐
29% total population☐
Density 27.3/km²☐
GDP - 13.8% of total

North

Manaus

North☐ East

Recife

Centre☐ West

Centre West☐
22% total area☐
6.4% total population☐
Density 5.9/km²☐
GDP - 5.3% of total

Brasilia

South☐ East

South East☐
11% total area☐
42.5% total population☐
Density 67.2/km²☐
GDP - 69.4% of total

Population density

Under 1/km²
1-9/km²
10-100/km²
over 100/km²

Rio de Janeiro

Sao☐ Paulo

South

South☐
7% total area☐
15% total population☐
Density 38.4/km²☐
GDP - 17.8% of total

Porto☐ Alegre

Figure 4.26 Brazil: relief, climate, agriculture and population

Factors involved in the incidence of disease

The incidence of disease is due to a combination of physical and human factors. These vary in importance throughout the world and account not only for the prevalence of disease but also for the different types of disease dominant in an area.

Physical factors

Climatic conditions have a considerable impact on the level of health. Hot, wet conditions encourage the spread of **infectious diseases** and provide ideal breeding conditions for insects which act as **vectors** (intermediate hosts which transmit disease but are not affected by it). Drought and arid conditions affect water supplies which in turn can result in increased levels of disease as water supplies become restricted and/or contaminated.

The quality of the water supply is frequently a determining factor in the level of health within a community. In the developing world, on average 50% of households lack access to clean water supplies, although this percentage rises dramatically in rural areas. In urban areas access to piped water supplies is easier to provide. In rural areas people may have access to a well deep enough to supply clean water all year round. More often, however, the well will be too shallow, liable to dry up in the dry season or become contaminated by sewage. For the poorest people, the water supply may be a considerable distance away and consist of a stagnant polluted river or pool. People, mainly women and children, have to make several daily trips on foot to carry this water back to their villages. This is time consuming and can have implications for the health of those who have to carry the water each day.

Related to the poor provision of clean water supplies is the lack of **basic sanitation** in the developing world. Few people, even in urban areas, have access to sewerage systems. As a result of this, diseases caused by contact with sewage are widespread in both urban and rural areas. Children are particularly at risk from diseases spread by contaminated water supplies. For example, 4 million children under five die of diarrhoea each year caused by having to drink contaminated water, despite the fact that the illness is preventable.

Mountainous areas usually have poor communications and are remote and isolated. In such communities, access to health care and medical services, even of the most basic kind, is difficult. As a consequence people's health levels are generally poor.

Human factors

A major cause of poor health and low levels of health care is poverty. Poverty means that many people live in overcrowded slum conditions lacking running water and sanitation. Others cannot afford housing and live on the streets. In such overcrowded, insanitary conditions disease spreads rapidly. Poverty leads to poor levels of education so that people are ignorant of the causes and treatments for even the most preventable diseases. Developing world countries are often unable to afford the mass immunisation programmes which have been so effective in reducing disease and death in the developed world.

Poverty also leads to poor levels of nutrition as people are unable to afford a healthy diet. Even in areas where people support themselves by farming, poor farming methods, overuse of the land and consequent soil erosion lead to yields that are too low to support the population. People who are undernourished are more susceptible to disease and once ill find it harder to fight their illness. As a result of all this, over one billion people (one-fifth of the total world population) lack adequate food, clean water, elementary education and basic health care.

World diseases

The lack of available clean water and adequate sewerage systems are major factors in the incidence of many of the most common diseases in the developing world. These can be classified as:

- **water related** – diseases transmitted via an insect vector breeding in the water, e.g. malaria and river blindness

- **water based** – diseases transmitted via parasites using hosts living in the water, e.g. schistosomiasis

- **water borne** – diseases transmitted by people using water contaminated by sewage, e.g. cholera and typhoid.

Malaria

Malaria affects 200–500 million people worldwide each year. In the past 15 years the disease has killed nearly 50 million people, with 2 million deaths in sub-Saharan Africa in 1999 alone. As many as 85% of the deaths occur in children, with one child dying every 20 seconds. Those who survive develop a limited measure of **immunity** in adulthood. They can however be reinfected, suffer from recurrent and violent bouts of fever, and are severely anaemic.

This weakens their immune system so that they are less resistant to other diseases. The disease also has adverse economic effects in areas where it is endemic (that is, part of the background to life). This is particularly damaging in areas where agriculture is a significant component of the GNP. It has been estimated that a single bout of fever causes the loss of ten working days.

We tend to think of malaria as a tropical disease associated with developing countries but this is not so. Apart from hot and cold deserts and high altitude areas, mosquitoes have a worldwide distribution (Figure 4.27). In general the incidence of the disease is found between 65 °N and 40 °S and below 3000 m. With global warming and rising average temperatures, disease-carrying mosquitoes may spread northwards into America and southern Europe.

The malarial **parasite** has a complex life cycle, as shown in Figure 4.28. In 1955, the **World Health Organisation (WHO)** spent hundreds of millions of dollars in an attempt to eradicate malaria. The disease was attacked on several fronts:

- First by the elimination of breeding areas. This was done by filling in and draining areas of stagnant water, spraying the areas with insecticides such as DDT and Malathion, spreading oil on the water surface, flushing streams to wash away the larvae and clearing trees and shrubs (which provide shelter for the mosquitoes) from streams and ponds.

- Second by preventing the development of the

Figure 4.27 Global distribution of malaria

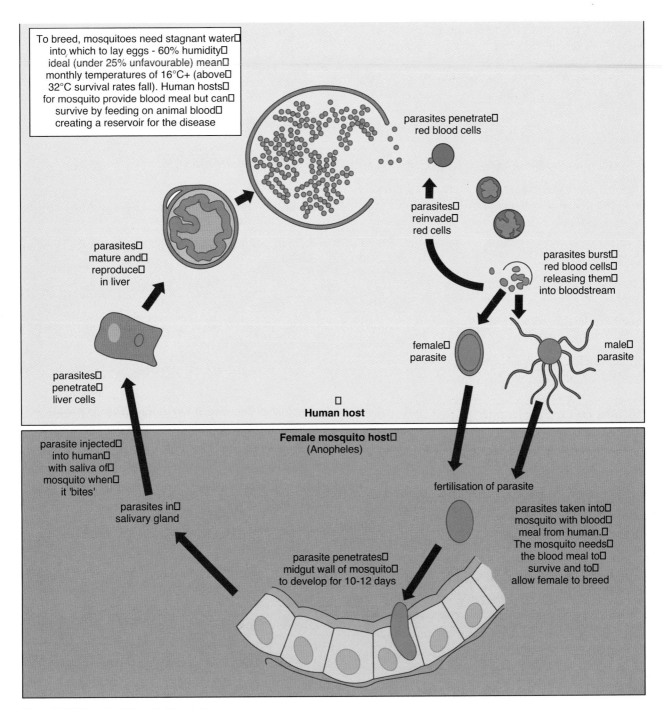

To breed, mosquitoes need stagnant water into which to lay eggs - 60% humidity ideal (under 25% unfavourable) mean monthly temperatures of 16°C+ (above 32°C survival rates fall). Human hosts for mosquito provide blood meal but can survive by feeding on animal blood creating a reservoir for the disease

parasites penetrate red blood cells

parasites reinvade red cells

parasites mature and reproduce in liver

parasites burst red blood cells releasing them into bloodstream

parasites penetrate liver cells

female parasite

male parasite

Human host

parasite injected into human with saliva of mosquito when it 'bites'

Female mosquito host (Anopheles)

parasites in salivary gland

fertilisation of parasite

parasite penetrates midgut wall of mosquito to develop for 10-12 days

parasites taken into mosquito with blood meal from human. The mosquito needs the blood meal to survive and to allow female to breed

Figure 4.28 Life cycle of the malarial parasite

larvae. This was done by introducing fish which eat the larvae, suffocating the larvae by spraying egg white on the breeding areas and using methods such as putting mustard seeds into paddy fields to drown the larvae by dragging them below the surface.

◆ Third by treating sufferers with drugs such as quinine to kill the parasites within their bloodstream, killing mosquitoes by spraying the interior of houses with insecticides such as DDT,

and running health education campaigns emphasising the ways in which the disease is transmitted and the need to seek early medical help, especially for children.

The effectiveness of these methods varied. The disease was eradicated from Europe, the former Soviet Union, most of the Middle East, North America and large areas of the Far East. Africa was virtually unaffected by the eradication methods and there was a resurgence of the disease in India in 1970. By 1969, the WHO realised that they could only exert some control over malaria, not eradicate it, and so abandoned their aim of worldwide eradication.

Many problems led to this situation. It proved extremely difficult to eliminate breeding areas. Mosquitoes need only a small area of water in which to breed, for example, in water that has collected in an old tin can or even in an animal hoofprint. Keeping an area clear of rubbish is not easy in urban slum areas; nor is the elimination of animal tracks in rural areas. The use of drug treatments is another problem area as the parasites are developing **resistance** to the present generation of antimalarial drugs, often as a result of the inappropriate use of these drugs in the past. In addition many countries are unable to finance a sustained antimalarial drug programme.

Despite the growing resistance to drugs it is still worthwhile treating the most vulnerable groups in endemic areas, such as pregnant women and people visiting from areas where malaria is not found. Mosquitoes are also becoming resistant to insecticides, and the use of the most common and effective insecticide DDT has greatly declined due to fears about its effects on the environment and the way in which it affects the long-term health of people and wildlife.

Despite these difficulties, doctors believed that by 1980, malaria had been eliminated from many areas. In recent years, however, there has been a resurgence of malaria worldwide. This is due to a combination of factors. A rise in average temperatures worldwide has allowed malarial-carrying mosquitoes to extend their

Figure 4.29 A Kenyan child suffering from malaria

range beyond the tropical and subtropical areas to which they had been largely confined. Global climate change has also had its effects. For example, in Kenya in 1998 there were four months of unusually heavy rain which created large areas of floodwater providing an ideal breeding ground. The result was a 500% rise in cases of malaria in the affected areas, with five children dying each day (Figure 4.29).

Another factor has been the collapse of public health programmes. For example, the break-up of the USSR means that the centrally controlled, large-scale antimalarial programmes of the past are no longer in place and malaria has reappeared in central Asia. Other health programmes in many developing countries have collapsed because debt repayments have meant that these countries are unable to maintain viable health services or to afford effective treatments. In other areas malaria has been spread because of a breakdown in civil government due to war, natural disasters and famine. These all lead to large-scale population movements. If these people are infected and move into an area that is free from malaria they will carry the disease into an area where people's immunity is non-existent, so that the disease can spread very rapidly with devastating effects on the victims.

The best advice at present is to try as far as possible to avoid being bitten. People should be especially careful at dawn and dusk when mosquitoes are most active. Precautions include having screens on doors and windows to prevent mosquitoes entering the house, using insecticides or fumigants to kill any that have penetrated into the house and sleeping under a net treated with insecticide. Drug treatment is still advised although it is losing its effectiveness.

In the long term the WHO would like to see the development of effective antimalarial drugs, in particular a **vaccine**. This is a difficult task as the malarial parasite is complex and is difficult to treat. The vaccine would have to be cheap, safe, easy to administer, be able to be used on infants and give lifelong immunity. It is extremely difficult to interest the private sector in such an expensive, technically difficult project with the prospect of poor returns on the finished product. Attempts are being made to speed up the process: for example, in 2003 Bill Gates donated $100 million for research, effectively doubling the amount of money available for malaria research.

Medical experts are now very hopeful that a major breakthrough can be developed and trials of a vaccine were carried out successfully in 2004. One of the new treatments being developed is based on a combination of currently available drugs and an extract from the Chinese herb *artemisin*. The main advantages of this new treatment are that it has a 95% success rate, the blood parasite has not developed a resistance to it, and it can be produced very cheaply, making it accessible to even the poorest families and individuals.

Schistosomiasis (Bilharzia)

It is estimated that this disease affects 200 million people worldwide (Figure 4.30) causing 20 000 deaths a year, mainly through kidney failure and bladder cancer. When the disease is untreated and becomes chronic it causes severe damage to the liver and the spleen. In areas with high incidence of the disease children are often particularly at risk. It is

found in over 70 countries, in tropical and subtropical areas, with almost half of these in Africa alone. It is one of the major communicable diseases of the developing world, probably the second most prevalent after malaria.

The disease is caused by a parasitic infection associated with a flatworm or blood fluke. The parasite has a complicated life cycle, as shown in Figure 4.31. It spreads when people come into contact with infected water while swimming, fishing, working on irrigated crops and whilst using the water for domestic or personal hygiene. People working in padi fields are particularly susceptible to schistosomiasis (Figure 4.32).

The situation is made worse by poverty, inadequate or complete lack of public health facilities, rapid urbanisation with the creation of slum conditions, and population movements such as those associated with refugees fleeing conflicts and other disasters. A further reason for the continuing spread of the disease has been the creation of large freshwater reservoirs in high-risk areas, such as the Aswan Dam in Egypt (Figure 4.33), the Akosombo Dam in

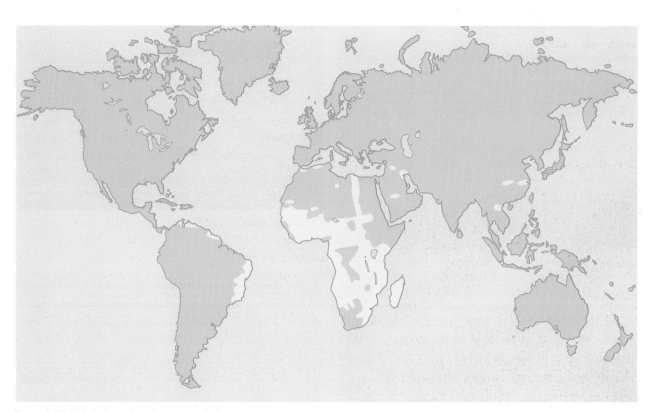

Figure 4.30 Global distribution of schistosomiasis

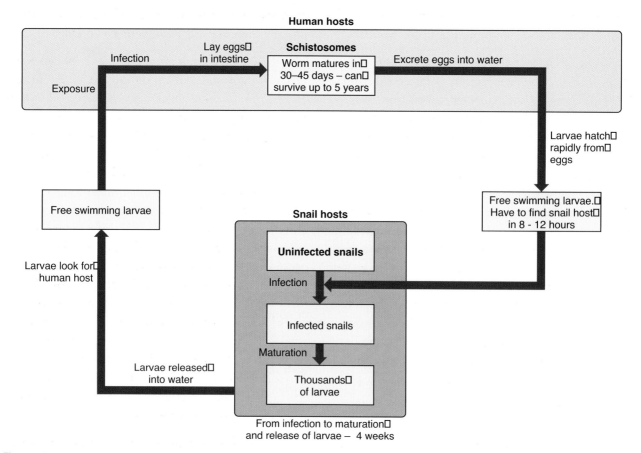

Human hosts

Infection — Lay eggs in intestine

Schistosomes
Worm matures in 30–45 days – can survive up to 5 years

Excrete eggs into water

Exposure

Larvae hatch rapidly from eggs

Free swimming larvae

Free swimming larvae. Have to find snail host in 8 - 12 hours

Snail hosts

Uninfected snails

Infection

Larvae look for human host

Infected snails

Maturation

Larvae released into water

Thousands of larvae

From infection to maturation and release of larvae – 4 weeks

Figure 4.31 Life cycle of the schistosomiasis parasite

Figure 4.32 Workers in a padi field are at risk from schistosomiasis

Figure 4.33 Aswan Dam in Egypt

Ghana, the Kainji Dam in Nigeria and the Kariba Dam in Zimbabwe. Associated with these large-scale projects there is usually an increase in irrigated land, creating large areas of slowly moving water which allow the parasites to spread outwards into environments that were previously hostile to them.

As the disease causes **chronic ill health** it has a high economic impact because it debilitates workers and affects children's growth and school performance. Its impact is, however, often unrecognised both by the public and by health authorities. Affected people do not see it as a serious health problem as its symptoms are often unspecific and take a considerable time to develop. This means that the disease often goes untreated in its early stages so that by the time treatment starts the symptoms can be very serious.

Schistosomiasis can be attacked on several fronts. People who are infected can be treated with drugs. These used to be dangerous with numerous side-effects. In recent years, however, new, safe and relatively cheap drugs have been developed. The problem for the developing countries is that, due to the large numbers of people infected, they cannot afford sufficient quantities of the drug and find distribution difficult. The host snails can be killed by using a **molluscicide** (chemical to kill snails) or by modifying the environmental conditions in which they breed, for example, lining irrigation ditches with concrete and trapping the snails in filters or wire mesh. The introduction of adequate sanitation systems and other health care measures, including health education warnings about the dangers of being in contact with infected water, would also make inroads into the incidence of the disease.

Despite the fact that diagnosis and control of schistosomiasis is relatively straightforward, interest in control measures has decreased over the years so that the disease is once again spreading, even in areas such as Brazil and China where effective measures in the past had almost eliminated the occurrence of the disease. At present there is no vaccine against the parasite, so the priority is to educate people about the dangers to which they, and especially their children, are exposed in high-risk areas.

Cholera

Cholera is believed to have originated in the Indian subcontinent and spread outwards as world trade routes developed. It reached Europe in the nineteenth century, causing many deaths. By the end of the nineteenth century a better understanding of the causes of the disease and improved water and sanitation systems, meant that it had once again retreated to the Indian subcontinent and south-east Asia. In 1961 a new strain of the cholera bacteria emerged in south-east Asia (the El Tor strain) and spread outwards from there, probably carried by Muslim pilgrims to Mecca. At that stage the spread was relatively slow. In 1991, however, there was a resurgence of cholera in South America, the first

epidemic for almost a century, which between 1991 and 1995 infected 1 million people and killed 11 000.

The cholera bacteria is spread via water contaminated by human waste containing the bacteria. It has long been associated with unsanitary conditions – a lack of adequate supplies of clean water and effective sewerage systems in overcrowded urban areas and in poor rural areas. Such conditions are always made worse by natural disasters such as flooding or any circumstances which force large numbers of people into overcrowded unhygienic conditions such as refugee camps. Many people in these circumstances are forced to use contaminated water supplies for cooking, drinking and washing food. Cholera also has traditional links with the sea. The cholera organism thrives best in moderately salty water such as coastal estuaries and it has been found that it can tolerate open ocean conditions by surviving in some species of plankton. It is only found in fresh water where nutrient levels from organic pollution are high. Throughout history epidemics have started in coastal cities with unhygienic conditions.

Part of the blame for the recent resurgence in cholera is due to deteriorating water and sanitation systems in areas where governments do not have sufficient money to maintain them. Many developing world cities have poor living conditions and inadequate housing due to rapid population growth. Many people suffer from malnutrition and there is the added problem of street vendors who do not use refrigeration or clean water. In the countryside there is an increasing use of waste water for irrigation. None of this, however, is sufficient to explain the worldwide re-emergence of cholera.

Several theories have been advanced to explain the latest epidemics. Was the bacteria carried around the world in the ballast and sewage of cargo ships and discharged into coastal waters? Was the bacteria simply dormant in coastal waters waiting for favourable conditions? Was there a dormant reservoir in the oceans waiting until the conditions for expansive growth were present? The discovery that the cholera organism could survive in some species of plankton does suggest a possible route by which cholera could spread rapidly over a wide geographical area carried by ocean currents.

Changing sea conditions such as the warming of surface waters by a change in the course of the El Niño current could allow an explosion of plankton numbers (called a **plankton bloom**) in coastal waters, and this, combined with nutrients from sewage, could re-awaken the cholera bacteria.

Cholera is treatable by drugs and oral rehydration therapy (replacing liquid and salts lost through sickness and diarrhoea). If it remains untreated, death is common due to acute dehydration, with children particularly at risk. The most effective way to control the disease is to provide safe, clean, piped and chlorinated water, and where that is not possible to dig deeper wells and introduce people to simple water purification techniques. The provision of effective sanitation systems is also vital to prevent contamination of water supplies. The root causes of overcrowding and poverty also have to be tackled in order for effective control measures to be put into place and maintained over the long term.

In the past people who had recovered from an attack of cholera were considered to be immune. However, the bacteria has since mutated and these people are now also at risk from re-infection. It makes the development of an effective vaccine unlikely, so control measures become even more important. The fact that plankton blooms may be implicated in the spread of cholera is also of importance. Plankton blooms are increasing in frequency as global temperatures rise, nutrient levels increase due to larger populations and atmospheric carbon dioxide levels rise. It could be possible to set up early warning systems so that people in coastal areas could take extra precautions, such as the chlorination of water supplies, if a plankton bloom approached. It is unlikely, however, that much progress will be made in dealing with cholera until there is an improvement in the supply of water and sanitation worldwide.

The provision of primary health care

Throughout the developing world, governments are unable to afford the elaborate health care systems found in the wealthier parts of the world. In an effort to improve the health of their citizens they have instead adopted a number of strategies classified as **primary health care**. This can be thought of as providing basic health care at the level at which it is most needed, for example, advice on diet, food preparation, dispensing of commonly required drugs such as anti-malarials and the treatment of minor ailments. It does not include such things as complex surgical procedures or treatments involving high-tech medical equipment.

One of the earliest attempts to provide basic health care to large numbers of rural dwellers was set up in China in the 1960s. The communist government was faced with a countryside that had virtually no access to health care. Most villages lacked medical facilities and the nearest health centres could be up to three days' walk away. Even if they could be reached, patients would find that the centres were poorly equipped and run by inadequately trained staff.

To deal with this situation large numbers of local people were trained as medical auxiliaries called **barefoot doctors**. They were given basic medical training which allowed them to treat straightforward illnesses and to refer more serious cases on to better-equipped hospitals. The typical **medical auxiliary** was a part-time health worker. The rest of the time they supported themselves in jobs such as farming. The costs of running the medical service were met by the community and a combination of Western and traditional Chinese medical practices were used.

In the early years the system was very successful at making inroads into the poor state of health in the countryside. In recent years, however, the service has faced considerable pressures. Despite the numbers trained it proved impossible to provide enough health care workers to satisfy the needs of rural communities. In addition, many were unable to continue paying for the training of auxiliaries and the upkeep of the health centres, so the system began to break down. To try to deal with this, the Chinese government started to encourage medical auxiliaries, traditional doctors and hospital physicians to set up in private practice. However, the introduction of private practice has not been to the benefit of large numbers of people. The countryside is generally too poor to attract health workers, who prefer to settle in the wealthier suburban areas. Once again many rural communities are suffering from a lack of effective, affordable health care. Unless the Chinese government can find the solution to these problems the situation is likely to deteriorate.

Despite this failure, the pattern has been adopted by many countries in the developing world as they struggle to improve the health of those living in rural areas. Many of these schemes have also run into difficulties. Burdened with debt repayments and forced to make structural readjustments by the World Bank, many countries are unable to fund their health services. In Tanzania, for example, medical treatment is no longer free and the ordinary citizen cannot afford to pay for the treatments. In other areas such as Mozambique, civil war and natural disasters destroyed what had been a highly effective health provision. Many schemes, however, do still exist, often funded by charities such as Christian Aid.

The charities tend to work with local organisations which are aware of local needs. In India, for example, Christian Aid works with local groups to train health volunteers and equip them with basic medical kits which are simple and cheap to provide and re-equip. The health volunteers are local people and so they are known and trusted by the communities which they serve. An integral part of this partnership is an emphasis on health education. The method is still an extremely cost-effective method of reaching large numbers of people in the countryside, allowing them to build a better future for themselves and their children.

The other main emphasis is to concentrate on low-cost, low-tech solutions which encourage self-help and cut reliance on the often expensive foreign medicines. Attempts have been made to identify medicines which can be produced locally, often based

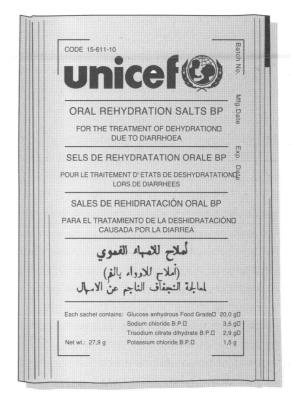

Figure 4.34 Oral rehydration salts for the treatment of diarrhoea

on traditional practices which have proved effective in the past. An example is the use of **oral rehydration therapy** (a solution of water, sugar and salt) to deal with the dehydration resulting from acute diarrhoea – one of the major causes of infant mortality in the developing world (Figure 4.34).

Although more expensive, the establishment of clinics to provide advice on nutrition, family planning and health education will bring benefits in the long term, as will large-scale child immunisation programmes which usually need foreign funding in order to be effective. All these measures, including foreign intervention, are needed if inroads are to be made into the poor levels of health provision throught the developing world.

CASE STUDY
Chogoria Hospital – Kenya

Chogoria Hospital is located 190 km north of Nairobi on the eastern slopes of Mount Kenya (Figures 4.35 and 4.36) at an altitude of 1540 m. The area around Mount Kenya at 2000 m is well watered (2000 mm of rainfall per annum) and has rich volcanic soils. Further east near the Tana River the land, at a height of 300 m, is flat and arid. In the well-watered areas farmers grow maize, millet, potatoes, fruit and vegetables for their own use and tea, coffee and pineapples as cash crops for export. In the driest areas people rely on subsistence farming and pastoralism and are generally much poorer with greater health problems. There are few tarmaced

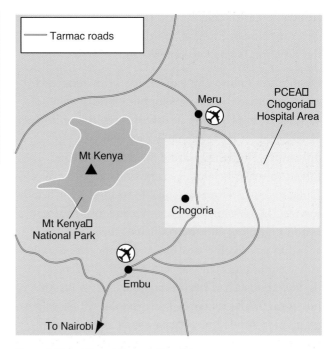

Figure 4.36 Location of Chogoria Hospital

Figure 4.37 In the rainy season, patients may find it difficult to reach the hospital

roads – most are dirt roads which are difficult to pass in the wet season (Figure 4.37). The hospital serves approximately 555 000 people with a population increase of over 2% per annum.

The hospital was established in 1922 as a Church of Scotland mission hospital and in 1923 treated 7000 outpatients (Figure 4.38). It grew steadily through the 1920s and 1930s until by 1936 it treated 760 inpatients and 11 606 outpatients. Numbers fell

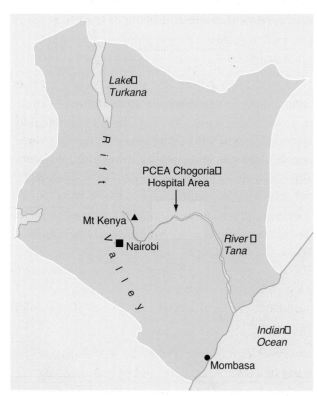

Figure 4.35 Location of Chogoria Hospital in Kenya

Figure 4.38 The clinic post where Chogoria Hospital started in 1922

Figure 4.41 Paediatric ward, Chogoria Hospital

affected the work of the hospital but work continued to expand, and once the emergency was over, there were 312 beds available for inpatients as well as a range of medical services. These included a surgical ward, maternity ward, TB treatment, dental and eye clinics. The hospital was also a centre for training nurses and midwives (Figures 4.39–4.41).

After Kenya gained independence the hospital continued its work in conjunction with the Kenyan government. By this time the hospital was beginning to experience considerable financial problems.

Figure 4.39 Surgical ward at Chogoria Hospital

during the Second World War as it was affected by shortages of personnel and equipment.

Expansion continued after the war with a greater range of services on offer. The Mau Mau uprising

From the start of the hospital's work there had always been a policy of charging a small fee, although no one in need and unable to pay was ever turned away. The problems became more severe due to several factors. The Kenyan government had promised free health care which in the event it was unable to deliver, except to outpatients and children. The financial problems became acute in the 1980s and 1990s as the country was suffering from the effects of debt repayment and the falling world prices for its major exports. In 1993 inflation was 100% and the hospital was forced to increase its fees by 50%. This has led to a steady decline in the numbers of patients treated (see Figure 4.42) and bed occupancy which had been as high as 160% in 1970 had fallen to 65% in 1999.

Figure 4.40 Mural on the wall of the community health building at Chogoria Hospital

The need for outreach into the surrounding rural areas has been seen from the beginning and by 1970 there were eight dispensaries staffed by past, leaving only the more expensive drugs as possible treatments. There has also been a worrying increase in the cases of TB and the problem of HIV and AIDS is on the increase. The greatest concern is the fact that an increasing proportion of the population can no longer afford hospital treatment and are either going without treatment or turning to unqualified healers.

In summary, the main reasons why primary health care is central to the services provided by Chogoria Hospital are as follows:

Physical

◆ Geographical isolation of the rural population means that many people cannot travel to the nearest hospital. As an example, one of Chogoria's associated clinics at Tharaka is 2–3 hours from the main hospital – or even longer during the rains.

◆ The 32 clinics operated by Chogoria are able to cover a much wider area, using staff effectively and efficiently. Specialist medical staff are able to visit the clinics on rotation rather than expecting all patients to travel to the hospital.

Social

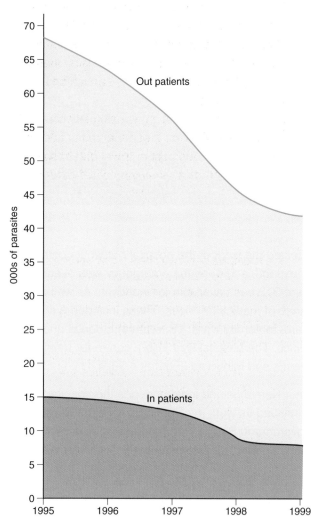

Figure 4.42 Number of patients being treated at Chogoria Hospital

hospital-trained nurses running day clinics. This network has gradually expanded until today there are now 32 clinics run by area health committees and offering a range of services. The clinics treat those who come in with diseases such as TB and malaria. They also undertake preventative work, for example, via a child immunisation programme. Ante- and post-natal advice is on offer as well as midwifery services. There is also a range of health education programmes including family planning advice and information on how to prevent the spread of HIV and AIDS. Once again attendance at these clinics is on the decline (Figure 4.43).

The major illness is still malaria, which is becoming resistant to the less expensive drugs used in the

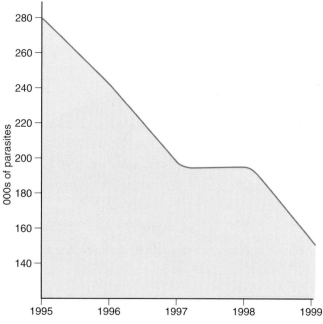

Figure 4.43 Decline in the number of patients treated in rural clinics

◆ Staff are local so there is less suspicion, closer contacts and fewer language problems.

◆ The system allows the development of a community approach to tackling health issues, for example the use of drama to educate people about HIV and AIDS.

Economic

◆ This system is much more cost-effective than large-scale projects.

◆ Since 80% of illnesses are water-related, basic health education through primary health care can eliminate many of these cases and reduce the number of 'unnecessary' hospital visits, thereby allowing the staff to treat more serious and specialist cases.

◆ Economic problems such as rising inflation or the market price of coffee mean that many people could not afford the cost of travel and hospital treatment, and as a result they could otherwise

QUESTIONS

1 Describe the physical and human factors which affect the incidence of diseases throughout the world.

2 For each of the diseases malaria, schistosomiasis and cholera:
  a describe the factors which affect its incidence and transmission
  b describe the effects on the economies of the areas involved
  c describe the measures taken to control and/or eradicate the disease
  d comment on the effectiveness of these methods.

3 Explain what is meant by primary health care and explain why it has been adopted as a health care policy by many developing countries.

4 Give examples of primary health care and comment on the appropriateness of such schemes in the context of the developing world.

5 Referring to specific examples from the text, identify the strengths and the problems of some primary health care policies.

6 The links between the health and development of a country are clearly illustrated by Figures 4.44 and 4.45. Using these statistics:
  a Describe and explain in detail the potential socio-economic impacts which HIV and AIDS may have on the development and health of sub-Saharan African countries, and on South Africa in particular.
  b Suggest strategies for overcoming these potentially catastrophic consequences.

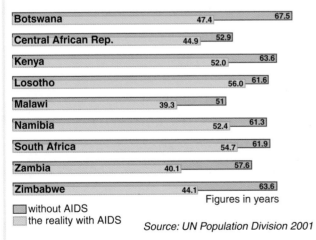

Figure 4.44 Potential impact of HIV/AIDS on life expectancy in sub-Saharan Africa

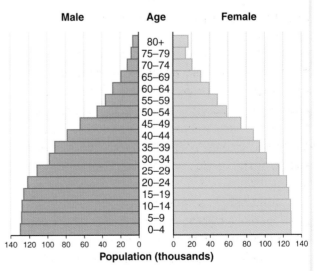

Figure 4.45 Potential impact of HIV/AIDS on South Africa's population structure

Key terms and concepts

- adjusted income per capita (p. 165)
- adult literacy (p. 164)
- barefoot doctor (p. 188)
- basic sanitation (p. 181)
- chronic ill health (p. 186)
- composite indicators of development (p. 164)
- cronieism (p. 170)
- Development Gap (p. 163)
- drug resistance (p. 184)
- educational attainment (p. 165)
- epidemic (p. 187)
- First World, Second World, Third World (p. 165)
- G8 countries (p. 174)
- Gross Domestic Product (GDP) (p. 169)
- Gross National Product (GNP) (p. 162)
- Human Development Index (HDI) (p. 164)
- immunity (p. 182)
- import substitution (p. 173)
- indicators of development (p. 162)
- infant mortality rate (p. 164)
- infectious disease (p. 181)
- infrastructure (p. 174)
- International Monetary Fund (IMF) (p. 173)
- life expectancy at birth (p. 164)
- literacy rates (p. 164)
- medical auxiliary (p. 188)
- molluscicide (p. 187)
- Newly Industrialised Country (NIC) (p. 167)
- North–South Divide (p. 165)
- oral rehydration therapy (p. 189)
- parasite (p. 182)
- Physical Quality of Life Index (PQLI) (p. 164)
- plankton bloom (p. 188)
- primary health care (p. 188)
- relative purchasing power (p. 163)
- Structural Adjustment Programmes (p. 173)
- subsistence activity (p. 163)
- sustainable growth (p. 162)
- Tiger Economy (p. 167)
- transnational corporation (p. 169)
- vaccine (p. 184)
- vectors (p. 181)
- World Bank (p. 173)
- World Health Organisation (WHO) (p. 182)

SUMMARY

Having worked through this chapter, you should now know:

- ★ there are marked inequalities in social and economic development globally, regionally and nationally
- ★ levels of development can be measured by a range of indicators – economic, social and composite
- ★ within the developing world, there are considerable differences in levels of development both between and within countries
- ★ the incidence of disease throughout the world is due to a combination of physical and human factors which include climate, water supply, levels of wealth and nutrition
- ★ malaria, schistosomiasis (bilharzia) and cholera each have distinctive features relating to their methods of transmission and distribution
- ★ there are a variety of strategies in place to control disease and its impact – these have had varying degrees of success.

The Examination
(PAPER 2: ENVIRONMENTAL INTERACTIONS)

THE STRUCTURE OF THE EXAMINATION

For the final examination in Higher Geography there are two papers:

◆ **Paper 1** covers the eight topics from Physical and Human Environments

◆ **Paper 2** covers Environmental Interactions.

The structure of Paper 2

In all there are six topics for Environmental Interactions, four of which are covered in this textbook. The question paper will have one question for each of the six topics and will be divided into two sections:

◆ **Section 1** Rural Land Resources, Rural Land Degradation and River Basin Management★

◆ **Section 2** Urban Change and its Management, European Regional Inequalities★, and Development and Health

★denotes topics not covered in this textbook

The examination for Paper 2 lasts for 1 hour 15 minutes and you will be required to answer **two** questions: **one** from **Section 1**, and **one** from **Section 2**.

Questions will be divided into a number of parts but most parts will generally carry more marks than those in Paper 1 (Physical and Human Environments) and will therefore require longer and more detailed answers.

Choosing your questions

During your Higher Geography course you will have studied at least two of the Environmental Interactions topics and these are the questions you should attempt – but ensure that you are clear about which they are. Do *not* try to answer questions from the other topics because you will not have the background knowledge to be able to answer these in the detail that will be expected. Make sure that you attempt

one question from each of the two sections of the paper.

Answering the questions

The examiners are not trying to catch you out, but rather will award marks when you give **specific and direct** answers to the questions which are set.

◆ Ensure that you read the question properly and that you tailor your answer accordingly. As well as developing the main theme of the question in your answer, make sure that you have underlined the **key instructions**. These are usually:

◆ *account for* – explain the cause of

◆ *analyse* – look at the issues, break them down and show how they interrelate

◆ *annotate* – add labels to show key features

◆ *assess* – weigh up the importance of

◆ *comment* – write an explanation

◆ *compare* – point out similarities and differences

◆ *contrast* – emphasise the differences only

◆ *describe* – state the main characteristics or features

◆ *discuss* – present arguments for and against an issue

◆ *evaluate* – review whether something has been successful

◆ *explain* – give reasons for

◆ *illustrate* – give an example, which could be a case study, sketch map or diagram

◆ *outline* – give the main features

◆ *suggest* – put forward a reason or an idea

◆ Although time is short in the examination, a very brief **outline plan of your answer** should take no more than a few minutes and will help you structure your answer.

◆ Make sure that you **answer all parts of the question**. Some may require a number of separate approaches within a single answer, for example, **'referring to social, economic and environmental factors …'**.

◆ Make good use of the case study material that has been included as part of your course. In this paper your ability to use relevant case studies is very important. Answers that are entirely abstract and do not use real world examples will not score highly.

◆ Use any resource material that has been provided as part of the question. Although at first sight some graphs or diagrams may look complex, in fact they are often very useful in providing you with broad headings to include in your answer.

Rural Land Resources

1

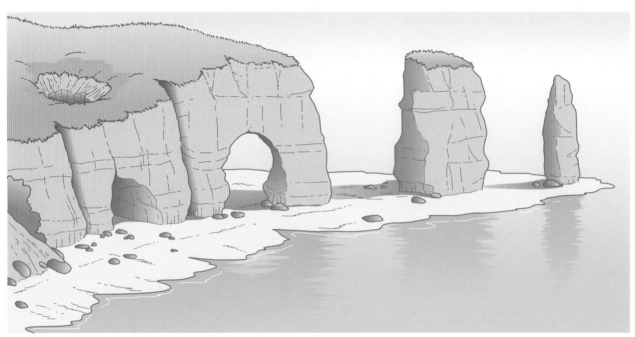

Reference Diagram Q1a Coastal landscape features

a Study Reference Diagram Q1a. With the aid of annotated diagrams, **describe** the main features of the physical landscape shown in Reference Diagram Q1a and **explain** the processes involved in their formation. 10

b For a named coastal area that you have studied, **explain** the **economic** and **social** opportunities provided by the landscape. 5

c *'One of the roles of the Pembrokeshire Coast National Park (Reference Diagram Q1b) is to help manage conflict between the local residents of Pembrokeshire and visitors to the area.'*

For Pembrokeshire or **any named** coastal area that you have studied:
i   give examples of environmental conflicts that have arisen, and
ii describe some of the measures taken to resolve these conflicts and comment on their effectiveness. 10

(25)

Reference Diagram Q1b The Pembrokeshire Coast National Park

2 a Study Reference Diagram Q2 which shows part of the glaciated landscape of the Cairngorm Mountains. With the aid of **annotated** diagrams explain how the main features of such a landscape were formed. 10

b With reference to the Cairngorms, or any other UK upland area which you have studied:

 i explain the social and economic opportunities created by the landscape

 ii give examples of environmental conflicts which may arise in the chosen area. 10

c Outline the arguments for and against the creation of National Parks in Scotland. 5

(25)

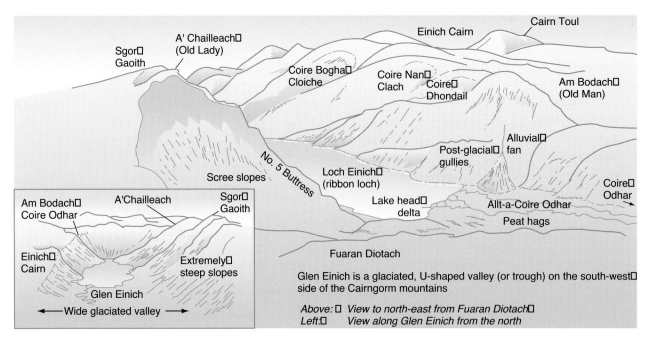

Reference Diagram Q2 Glen Einich – a glaciated valley

Rural Land Degradation

3 a Study Reference Diagram Q3.

For **either** the Great Plains or the Savanna Grasslands, and referring to specific examples:

i describe the human activities which have resulted in land degradation 8

ii explain in detail the effects of the removal of vegetation cover on both land and people 8

b Choosing specific examples from North America and Africa north of the Equator, describe the effectiveness of measures taken to try to conserve the soil in these areas. 9

(25)

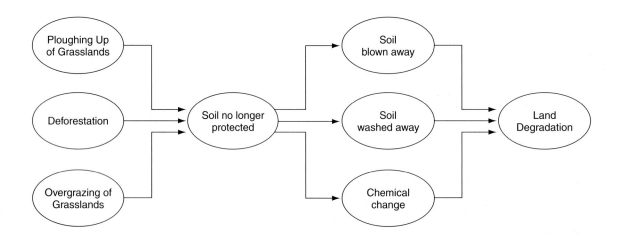

Reference Diagram Q3 The processes involved in land degradation

4

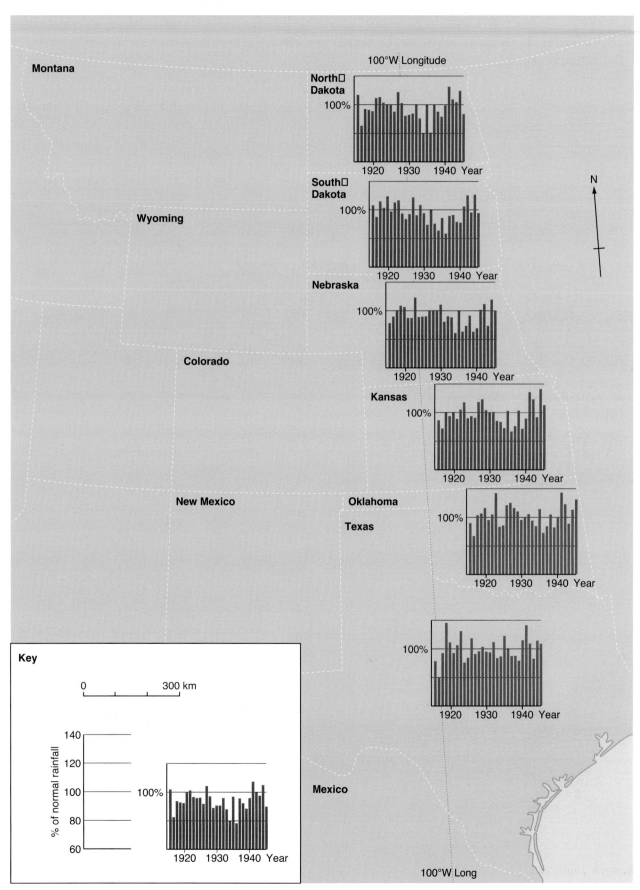

Reference Diagram Q4a Rainfall patterns in the Great Plains

a Study Reference Diagram Q4a.

Describe the annual rainfall patterns and show how variations in rainfall may lead to land degradation. 5

b For either the Great Plains of USA or the African Sahel describe the human factors which have led to rural land degradation. 10

c Reference Diagram Q4b illustrates the work of the Tennessee Valley Authority in the USA. Choose any four methods of soil conservation shown in the diagram and explain how each helps to conserve soil and reduce land degradation. 10

(25)

1 reforested land
2 gully erosion halted by check dams, trees planted on gully banks
3 contour cultivation practised on lower land
4 as erosion stops, flooding is reduced
5 new reservoir supplies power to nearby villages
6 river navigation improves, fish catch increases
7 urban services improve as population migration stops
8 shelter belts reduce wind erosion, pastures are improved or upgraded
9 absence of landslides and floods improve road communication
10 rural services improve as villages expand
11 crop rotation practised in strips along contours
12 forested slopes prevent siltation of reservoirs

Reference Diagram Q4b The work of the Tennessee Valley Authority

Urban Change and its Management

5

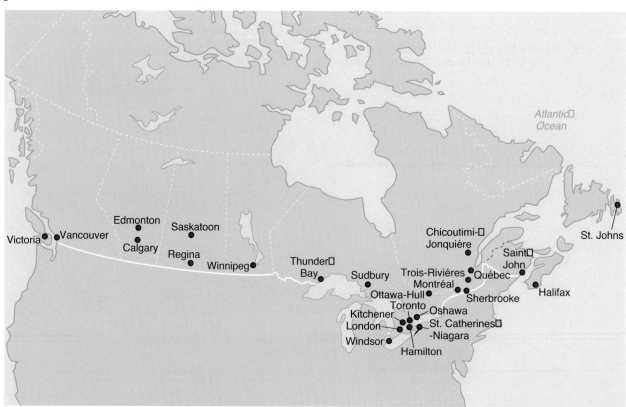

Reference Diagram Q5a The distribution of major Canadian cities

a Study Reference Diagram Q5a.

Describe and explain the distribution of major cities in Canada, or in another **developed** country you have studied. 5

b Study Reference Diagram Q5b.

Apply the models of processes influencing the changing location of urban land uses to a city from the **developed world** which you have studied, and comment on any similarities and differences. 10

c With reference to a city which you have studied in the **developing world**:
   i describe and account for its rapid growth
   ii identify the social, economic and environmental problems which have resulted from its rapid growth. 10

(25)

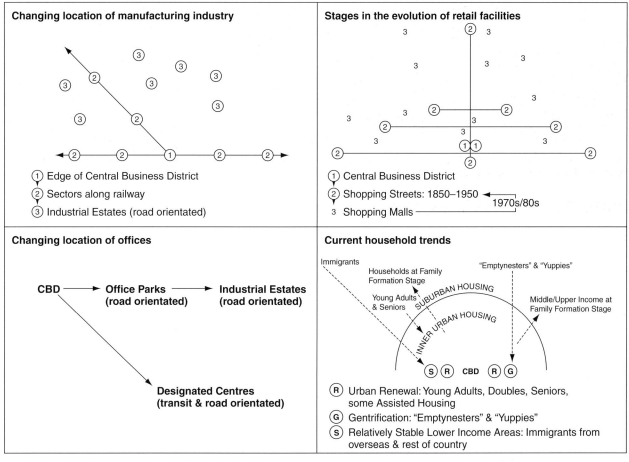

Reference Diagram Q5b Processes influencing urban land use change

6

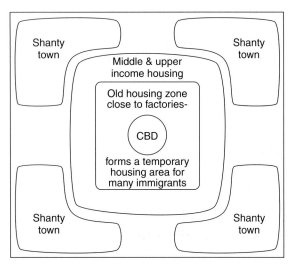

Reference Diagram Q6 Model of a city in the developing world

a Study Reference Diagram Q6.
   Compare and contrast the land use pattern of a city which you have studied from the **developing world** with the model, highlighting similarities and differences. 5

b With reference to any named city from the **developing world**:
   i describe the main problems associated with shanty towns
   ii examine the solutions proposed, and comment on their effectiveness. 10

c Referring in detail to a named city in the **developed world**, describe and explain the changes which have taken place in housing, industry, shopping and transport since 1950. 10

(25)

Development and Health

7 a Describe and explain the differences which may exist in levels of development **within** countries in the **developing world**. Your answer should refer to at least one specific country. 5

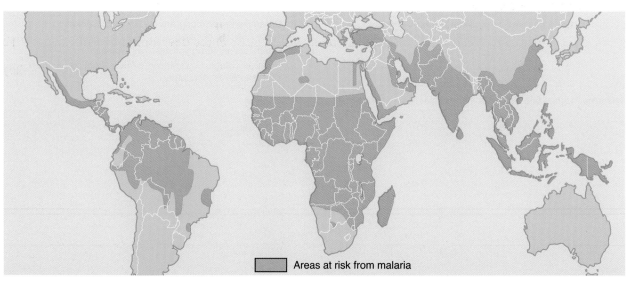

Areas at risk from malaria

Reference Diagram Q7a Areas at risk from malaria

Reference Diagram Q7b Resistance to either Chloroquine or Fansidar

b Study Reference Diagrams Q7a and Q7b.
Choosing either malaria or another named disease in the **developing world**:
    i   outline the physical and human factors which contribute to the spread of the disease
    ii describe methods used to control the disease and comment on their success.   12

c Using specific examples of primary health care, explain why it is an appropriate policy for improving health care provision in many **developing countries**.   8

                                                                 (25)

8 a Give **three** examples of indicators which might be used to illustrate a country's level of development, and for each comment on their value for this purpose.   6

b **i**   explain why such indicators may not provide an accurate picture of the true standards of living **within** a country.
    ii Give reasons for the differences in levels of developments which exist **between** different countries in the **developing world**.   7

c Referring to specific examples, outline the importance of:
    i   primary health care
    ii international organisations
        in attempting to combat the effects of any **one** named disease in the **developing world**.   12

                                                                 (25)

INDEX

A

abrasion 8
access to countryside (UK) 5–6
adjusted income per capita 165
adult literacy 164
aeolian erosion 91
afforestation 93–4, 103–4
Africa, causes of problems 173–4
agroforestry 103–4
arable farming 98–9
Asian Tigers 169–70
attrition 8
autonomous authority 148

B

backwash 8
barefoot doctors 188
basic sanitation 181
beach nourishment 25
Beauly to Denny transmission line 64–5
bilharzia *see* schistosomiasis
biodiversity 5, 60
biological action, soil erosion 70
bison ranching 95–6
Blue Box System 140
bluff 119
Bonington, Chris 57
boom irrigation 96
braided river channel 51
British Raj 142
brownfield site 123
budgeting 147
Burkina Faso case study 105–7

C

Cairngorm mountains 48–52
    conservation issues 53–5
    expansion of skiing 55–7
    forestry 60
    plateau edges and valley 49–52
    plateau surface 48–9
Cairngorm National Park 53, 54
Canada
    population growth 114–40
        urban trends 117–19
        urbanisation 114–17
candidate Special Areas of Conservation (cSAC) 55
carboniferous limestone 27
    landscapes 27–44
cash crops 101
cell grazing 95
Census Metropolitan Area (CMA) 115
Central Business District (CBD) 119, 127–9, 131–2
centre-pivot irrigation 96
chemical erosion 70–1
chernozem 89–90
China, health care 188–9
Chogoria Hospital, Kenya 190–3
cholera 187–8
chronic ill health 186
clearances 58
climate 89, 97–8, 168
clint 30
coastal erosion
    hard defence 24, 25–6
    soft defence 24–5
coastal landscapes (UK) 8–26
    processes 8–10
coasts, concordant 10
    discordant 10
colonial era 142
Common Agricultural Policy 16
Community Woodland Supplement 60
commuting 137
composite indicators of development 164–5
condominium 130, 133
conservation 19, 52–5
    of people in villages 58–60
    soil 70, 85–6, 95
constructive waves 8, 9
contour ploughing 85, 94
core and periphery 134
corries 48–51
corrosion 13
Countryside Access and Activities Network (CAAN)
    (Northern Ireland) 6
Countryside Agency 36
Countryside and Rights of Way (CRoW) Act (2000) 5–6
cronieism 170
crop changes 86
crop rotation 86
currency values 163

D

dale 27
decentralisation 129
deforestation 100–1
deindustrialisation 132–3
deposition 3, 4, 9, 72
desertification 97, 102, 105
destructive waves 8
detachment 72
development 162–80
    contrasts in 167–9
    defining 162–7
    factors influencing 168
    indicators 162–5
    measuring 165–7
Development Gap 163
diguettes *see* stone lines
diseases
    factors in incidence 181
    world 181–8
doline 30
Dorset
    coastal defence 24–6
    coastline 18–24
    impact of the Olympics on 22–3
    Lulworth Cove 10, 21–2
Dorset and East Devon Coast World Heritage Site 19
double cropping 86
drip irrigation 105
drug resistance 184
dry valley 29
dust bowl 93

E

Earth Heritage 55
economic growth 162
edge cities 129–30
education
    on land degradation 96, 105
    access to 103
educational attainment 165
electricity supply 152
England
    rights of way 5–6
    rural population 36–7
English Nature 31
Environment Act (1995) 27
environmental impact 61, 64
environmental management systems 39–40

Environmentally Sensitive Area (ESA) 33, 34–5
epidemic 187
erodibility 75
erosion 3, 4, 69
    coastal 8–9, 24–6
    fluvioglacial 45
    glacial 45–52
    linear 51
    soil 70–7
erosivity 75
ethnic groups, and inequalities 123–5
examination, answering questions 195–8

F

fallow period 95
farm diversification 32–4
farming 31–4, 81, 82, 85–6
    see also arable farming; pastoral farming
fertilisers, chemical 82–3, 86
fetch 8
fiord 10, 46
First World 165
Firth of Forth 11–14
    coastal features 12–13
    improvements in water pollution 14
    land use 13–14
    petrochemicals 13–14
fixation points 108
flood alleviation 24–5
flood control 86–7
flood irrigation 101
fluvioglacial erosion 45
footloose industries 37
forestry 60, 81, 82
Forth Estuary Forum 14
fuelwood 100–1

G

G8 countries 174
    population growth rates 115
Gates, Bill 184–5
gentrification 123
geology 3, 10, 12, 29, 38
glacial drift 30
glacial till plain 120
glaciated landscapes 45–66
glaciation 4, 45–6
glacier 45, 46

glaciokarst 27, 30
Glasgow 78, 115
global warming 17, 44, 62–3
GO train 138
gorge 29
government
    corruption in 168
    investment 168
    policy 60
    see also local government
Great Plains case study 89–96
    causes and consequences of rural land degradation 91–3
    climate 89
    human activity 90–1
    natural environment 89–90
    solutions to land degradation 93–6
Great Scar Limestone 29
Greater Toronto Area (GTA) 122, 140
"greying" population 123
gritstone 39
Gross Domestic Product (GDP) 169
Gross National Product (GNP) 162
groynes 25–6
gryke 30
gully 74, 102
gully erosion 72, 73–4

H

harmattan 98
health 181–94
Highlands (Scotland)
    and hydro-electricity 61–5
    rural depopulation 56, 58
hinterland 143
honeypot site 42
housing, development 154–5
Human Development Index (HDI) 164–5
Hutton, James 12
hydraulic action 8, 13
hydro-electric power (HEP) 60–4, 86–7
    "run-of-river schemes" 61–5
hydrogen 16–17

I

igneous rocks 2
immunity 182
import substitution 173
income loss to farmers 103

income per capita, adjusted 165
income polarisation 134
indicators of development 162–5
industrial development 87, 130–1
industrial suburbanisation 132–3
inequalities
    in development 163–70
    and ethnic groups 123–5
infant mortality rates 164
infectious diseases 181
infiltration rate 77
infrastructure 139–40, 148, 156–7, 174
inner city redevelopment 135
integrated transport network 138
intercropping 86, 95
interglacials 30, 45
International Monetary Fund (IMF) 173
Intertropical Convergence Zone (ITCZ) 98
Ireland, Northern and Southern, access to rural areas 6
irrigation
    boom or centre-pivot 96
    drip 105
    flood 101

J

jiko 105

K

Karachi
    case study 145–57
    improving housing 154–5
    improving infrastructure 156–7
    improving transportation 156
    managing change 147–8
    Orangi Pilot Project 152, 157
    physical growth 146–7
    role of authorities 147–8
    site, situation and structure 145–6
    squatter settlements 148–53
    urban differences 153–7
karst 27, 28, 29–31
katchi-abadis 143, 148–53
    development of 149–51
    employment in 153
    house types 151–2
    services 152
Kenya case study 174–80
    Chogoria Hospital 190–3

L

Laggan, conservation of people 58–60
land degradation 67–111
   defined 68–77
   in North Africa 97–111
   in North America 78–96
land management 5–6, 19, 70
land ownership 5–6
Land Reform (Scotland) Act (2003) 6, 16
landscape change, processes 3–4
landscape diversity 60
laser levelling 94
LEDC 173–80
   population growth and urbanisation 141–59
legislation (UK) 5–6
less economically developed country *see* LEDC
life expectancy at birth 161, 164, 165
Light Rapid Transit (LRT) 138–9
limestone pavement 30, 38, 39
limestone scar 30
linear erosion 51
literacy rates 164
Liveable Metropolis 134
local government, and urban change 125–7, 140, 147–8
Loch Lomond Re-advance 46
Loch Lomond and the Trossachs National Park 53
lochan 49
longshore drift 9, 13
Lulworth Cove, Dorset 10, 21–2

M

machair
   definitions 14
   South Uist 14–18
malaria 182–5
malnutrition 102
MEDC, population growth and urbanisation 114–40
medical auxiliary 188
metamorphic rocks 2
Metropolitan Area 145
migration 102
minimum tillage 94–5
mining 81, 82, 85
molluscicide 187
monoculture 82, 92, 101
more economically developed country *see* MEDC
mulching 95

N

National Nature Reserve (NNR) 42, 54, 55
National Parks 27, 52–3
National Parks and Access to the Countryside Act (1949) 27
National Scenic Areas (NSAs) 52–3, 55, 63–4
newly industrialised countries (NICs)167 168–73
Niger case study 107–9
nivation 49
nomadic people 99
North Africa, land degradation 97–111
North America, land degradation 78–96
North-South Divide 165, 166

O

Olympics (2012), impact on Dorset 22–3
open spaces 135–7
oral rehydration therapy 189
overcultivation 100
overdependency 102
overgrazing 71, 95, 100

P

Pakistan
   phases of development 142
   urban population growth 142–3
   urban trends 143
   urbanisation in 141–57
parasite 182
Partition 145
pastoral farming 95, 99, 104–5
periglacial conditions 29
permafrost 30, 45
permeable rock 27
physical erosion 70–1
Physical Quality of Life Index (PQLI) 164
plankton bloom 188
planning 147
plant cover *see* vegetation
ploughing 82
   minimal 94–5
   see also contour ploughing
political corruption or bias 168
population
   ageing 123
   growth in an LEDC 141–59
   growth in an MEDC 114–40
   growth and land degradation 69

rural in England 36–7
   rural-urban shift 5, 113
pothole 30
poverty 69, 181
PQLI (Physical Quality of Life Index) 164
primary health care 188–9, 192–3
public transport 137–8
public utilities 147

Q

quarrying 38–40

R

rainfall 76–7
rainsplash 71, 72–3
RAMSAR site 14, 54
ravine 120
reforestation 85
regional shopping malls 129
relative purchasing power 163
relief 76–7
relocation 132–3
renewable energy 16, 60–4
reurbanisation 135
rights of way 5–6
rill 73, 102
rill erosion 72, 73–4
river terrace 46
roches moutonnées 48
rock types 2–3, 27
rotational sliding 49
runoff 76, 86
rural depopulation 56, 58
rural land degradation 67–111
rural land resources 1–66
rural landscapes in the UK 2–7
   human activity 4–5
   land ownership, management and legislation 5–6
   physical factors 2–4
rural services, survey (1997) 34–6
Rural Stewardship Scheme 16
rural-urban contrast 168

S

Sahel case study 97–105
   causes and consequences of rural land degradation
   99–103

climate 97–8
   human activity 98–9
   natural environment 98
   solutions to land degradation 103–5
salinisation 71, 101, 102
saltation 75
sanitation, basic 156, 181
schistosomiasis 185–7
Scotland
   coastline 8
   conservation in mountain areas 52–5
   glacial landscapes and processes 46–7
   hydroelectricity in the Highlands 61–5
   life expectancy at birth 161
   National Parks 53
   rights of way 6
   see also Highlands (Scottish)
Scottish Forest Parks 60
Scottish Natural Heritage (SNH) 14, 53, 56, 63
Scottish Outdoor Access Code 6
scree 48
Second World 165
sedentarisation 108
sedimentary rocks 3
services 147, 152
   rural in UK 34–36
sewerage 152, 157, 181
shake hole 30
shanty housing 147
   see also katchi-abadis
sheet flow 73
sheet wash 71, 73
shelter belt 93–4
shopping plazas 129
Singapore case study 170–3
sink 30
site 119
Site of Special Scientific Interest (SSSI) 12, 14, 31, 54–5
slope stabilisation 135
soil conservation
   education in 96
   techniques 70, 85–6
Soil Conservation Service (SCS) 95, 96
soil degradation 68–70
soil erosion
   by water 71–4
   by wind 74–5, 91
   factors affecting 75–7
   processes in 70–2
   Tennessee Valley 84
soil exposure 83
soil formation 68–9

soil type, and structure 77
solifluction 49
solution 8
South Uist, machair 14–18
Special Areas of Conservation (SAC) 55
Special Protection Areas (SPAs) 12, 14, 54
spit 13, 119
squatter settlements *see* katchi-abadis
stalactite 30
stalagmite 30
standpipes 151
starvation 102
stone lines 104
storms, impact of severe 17–18
stoves, fuel-efficient 105
strip cropping 85–6
strip cultivation 95
strip-commercial 129
Structural Adjustment Programmes 173
subsistence activities 163
suburban business districts 129–30
suburbanisation, industrial 132–3
surface creep 75
surface runoff 76–7
suspension 74–5
sustainability 55
sustainable growth 162
swallow hole 30
Swanage, coastal protection 25–6
swash 8
sweep plough 94–5
Swinden Quarry 38–40

T

Tennessee Valley 78–89
    causes and consequences of land degradation 82–4
    climate 78
    flood control, HEP and industrial development 86–7
    land degradation 78–89
    land use 80–1
    social and economic conditions 81
    soils and vegetation 78–9
Tennessee Valley Authority (TVA) 85–8
Third World 167
tidal power 16
Tiger economies 167, 169–70
till plain 120
tillage, minimum 94–5
topsoil, loss of 102
tor 49

Toronto case study 119–40
    Census Metropolitan Area 116, 122
    Central Business District 127–9, 131–2
    deindustrialisation and relocation 132–3
    ethnic groups and inequalities 123–5
    housing 133–4
    industry 130–2
    infrastructure, water and waste 139–40
    inner city redevelopment: St Lawrence 135
    managing land use change 127–30
    managing urban change 125–7
    open spaces 135–7
    population distribution and growth 115, 116, 122–3
    retailing 129
    situation, size and growth 119–22
    suburban business districts 129–30
    transport 128, 137–9
tourism 40–3, 55–7, 58–60
and conflicts 21–2
transnational corporations 169
transport 128, 137–9, 152
    integrated network 138
transportation 3, 4, 9, 72, 156

U

U-shaped glacial trough 49
UK
    coastal landscapes 8–26
    rural landscapes 2–7
urban change and management 113–59
urban fabric 135
urban redevelopment 23
urban renewal 135
urban sprawl 137
urbanisation 101, 141
    in an LEDC 141–59
    in an MEDC 114–40
    "push" and "pull" factors 142–3

V

vaccine 184–5
vegetation 76
vertical development 129–30
villages, changes 34–7, 58–60
visual pollution 61

W

Wales, rights of way 5–6
waste disposal 139–40, 157
water pollution 14
water supply 139–40, 152, 156, 164, 181
waves 8–9
weathering 3, 4
Wester Ross, hydro-electricity in 63–4
Weymouth, Preston Beach Road flood alleviation 24–5
wilderness 54
wind power 16
Woodland Grant Scheme, Community Woodland
    Supplement 60
World Bank 173

world diseases 181–8
World Health Organisation (WHO) 182
World Heritage Site 18, 55

Y

Yorkshire Dales
    quarrying 38–40
    recreation and leisure 40–3
    rural landscapes and settlement patterns 31–7
Yorkshire Dales National Park 27–44
    conflicts 37–44
    landscape features and processes 29–31